1st International Conference on Applied Computing and Smart Cities (ICACS21)

AA004400

ITM Web of Conferences Volume 42 (2022)

Erbil, Iraq
22 – 23 January 2022

Editor:

Ali Kattan

ISBN: 978-1-7138-4453-2

Printed from e-media with permission by:

Curran Associates, Inc.
57 Morehouse Lane
Red Hook, NY 12571

proceedings.com

Some format issues inherent in the e-media version may also appear in this print version.

Printed with permission by Curran Associates, Inc. (2023)

For additional information, please contact EDP Sciences – Web of Conferences
at the address below.

EDP Sciences – Web of Conferences
17, Avenue du Hoggar
Parc d'Activité de Courtabœuf
BP 112
F-91944 Les Ulis Cedex A
France

Phone: +33 (0) 1 69 18 75 75
Fax: +33 (0) 1 69 28 84 91

contact-edps@webofconferences.org

Additional copies of this publication are available from:

Curran Associates, Inc.
57 Morehouse Lane
Red Hook, NY 12571 USA
Phone: 845-758-0400
Fax: 845-758-2633
Email: curran@proceedings.com
Web: www.proceedings.com

TABLE OF CONTENTS

Knowledge UNIVERSITY

ICACS21

1st International Conference of Applied Computing & Smart Cities

20-21 December 2021, Erbil, Iraq

https://sites.google.com/view/icacs21

ITM Web of Conferences

doi Clarivate ProQuest DOAJ DIRECTORY OF OPEN ACCESS JOURNALS IET Inspec 万方数据 WANFANG DATA

See https://www.webofconferences.org/itm-web-of-conferences-forthcoming-conferences

ICACS21

Cities stand to benefit the most from connecting people, process, data, and things. A Smart City exploits sustainable information and communication technologies to improve quality of life, health, education, and also the performance of urban services for citizens. Knowledge University invites you to submit proposals to ICACS21 to address these issues.

The conference will be held at Knowledge University campus in the city of Erbil, Iraq. However, due to the COVID19 restrictions remote participation via video conferencing is also possible.

Accepted papers will be published in a conference proceeding book with indexing by **ITM Web of Conferences** including **DOI** (Crossref), **Clarivate Analytics** (web of science), **DOAJ**, **ProQuest** and **Google Scholar** and others.

Topics include, but not limited to:

- AI and Machine Learning for secure and sustainable smart cities
- Cloud/edge/fog computing for smart cities
- IoT and Cyber Physical Systems for smart cities
- Secure and Trustworthy Computing for smart cities
- Energy harvesting and sustainability in smart cities
- Cloud/edge/fog computing for smart cities
- Employment of IoT and mobile network technologies in education
- AI enabled IoT education systems
- Machine learning for IoT Systems
- IoT for health care systems
- Sustainable power system based on IoT and AI
- Protocols and algorithms for wireless sensor networks
- Internet of Flying Things
- Unmanned Aerial Vehicles (UAVs) in IoT
- Social media and urban awareness
- Blockchain for IoT networks
- 5G and beyond 5G networks
- Wearable devices for communications and computing

A.K August 2021

Statement of Peer review

In submitting conference proceedings to ITM Web of Conferences, I certify to the Publisher that I adhere to the **Policy on Publishing Integrity** of the journal in order to safeguard good scientific practice in publishing.

1. All articles have been subjected to peer review administered by the proceedings editors.
2. Reviews have been conducted by expert referees, who have been requested to provide unbiased and constructive comments aimed, whenever possible, at improving the work.
3. Proceedings editors have taken all reasonable steps to ensure the quality of the materials they publish and their decision to accept or reject a paper for publication has been based only on the merits of the work and the relevance to the journal.

Title, date and place of the conference

1st International Conference on Applied Computing & Smart Cities (ICACS21)
22-23 January 2022, Erbil - Iraq

Proceedings editor(s):

Asst. Prof. Dr. Ali Kattan

Date and editor's signature

18 Feb 2022

ITM Web of Conferences **42**, 01001 (2022) https://doi.org/10.1051/itmconf/20224201001
ICACS21

Using Decision Tree Algorithms in Detecting Spam Emails Written in Malay: A Comparison Study

Saifuldeen H Abdulrahman [1*], and *Mohammad* Salim [2]

[1] Computer Science Department. College of Science, Knowledge University, 44001 Erbil, Kurdistan Region
[2] IT Department, Faculty of Science, Tishk International University, Erbil, Iraq

Abstract. Emails have become the most economical and fastest communication forms. However, during the past few years, the increment of email users has dramatically increased spam emails. Various anti-spam techniques have been developed to minimize if not eliminate the spam problem. In this paper, we study the disparity in the effectiveness of using different decision tree algorithms in email classification and combat spam problems. For that, we have chosen Universiti Utara Malaysia emails as a case study. To achieve the best possible classification accuracy, we compared all chosen algorithms' performance, which are Random Forest, LMT, Decision Stump, J48, Random Tree, and REP Tree. The experimental results showed that the Decision Stump algorithm is more effective to be used in classifying the emails, and the F-measures, Precision, and recall score for the Decision Stump algorithm are higher than the other comparison algorithms.

1 Introduction

Email is an electronic messaging system that sends messages through networks, allowing users to connect with one another at a low cost while also offering a reliable mail delivery system. Because of its dependability, user-friendliness, and vast range of free email, it is the most popular and preferred communication tool [1]. Email is a repository for messages, and the compose, send, receive, and store methods rely on electronic communication systems. Email management is a significant and growing issue for organizations and individuals because it is prone to mismanagement. Unfortunately, some issues with this service increase dramatically proportional to the development and spread of email services and the increased reliance on email by users [2].

The combination of low-cost, high-bandwidth Internet connections, falling storage costs per megabyte, and increased email users has resulted in an explosion of email data per user. Sorting through all this data and determining what is valuable and what is not is a daunting task. Spam is one of the most destructive problems facing email today. Spam, bulk email, or junk email are all terms that refer to inappropriate or irrelevant messages sent to many recipients via the Internet. Spam also refers to email that was not requested by the user [2, 3]. Typically, spam contains advertisements for dubious products and services such as "Make

* Saifuldeen H Abdulrahman : saifuldeen.abdulrahman@knu.edu.iq

Money Fast" schemes, multilevel marketing, illegally pirated software, and foreign bank scams. Spam may also contain offers to sell real estate, medicine, loans, and investments. Spam, or unsolicited email, is widely regarded as a serious threat to the Internet, as it floods users' inboxes and costs businesses billions of dollars in wasted bandwidth. Spam's global productivity cost increased by $2 billion to $132 billion in 2010 [4].

Additionally, spam causes a number of negative consequences, including an overflow of email storage capacity, which disables the server's ability to receive new emails, as well as a slow response time from the server and insufficient system resources. Additionally, email spam wastes the user's time in containing and deleting unwanted emails, results in the loss and/or delay of critical or emergency email messages and degrades Internet bandwidth and performance. Additionally, spam is one of the most effective methods of spreading malicious programs, worms, and Trojan Horses that corrupt computers and operating systems. Additionally, spam contributes to an increase in the proportion of people exposed to fraud, resulting in the loss of millions of dollars worldwide each year [4].

These and other negative consequences motivate researchers and businesses to work toward the elimination of spam and the abolition of harmful effects. Spam is available in a wide variety of languages, including Malay, Arabic, Korean, Chinese, and other Asian dialects, but is most frequently written in English. As a result, manually classifying too many emails is difficult, and machine learning techniques must be introduced. The machine learning technique entails developing filters to examine and eliminate characteristics from a corpus of spam [5]. Spam detection is not a typical text categorization task since it has some intriguing characteristics. Both spam and legitimate messages can cover a wide variety of topics and genres. In other words, neither class is homogenous. Additionally, email messages range in length from a few text lines to dozens of text lines. Moreover, the message may contain grammatical errors and unusual acronyms (sometimes intentionally used by spammers to fool anti-spam filters). As a result, the learning model should be robust under such circumstances.

This study aims to implement a variety of decision tree algorithm techniques for email classification and determine the most effective algorithms. Due to the growing number of documents, the variety and size of the dataset, and the variety of languages used to write an email. Email classification poses significant challenges due to the difficulty distinguishing spam from legitimate emails [6, 7]. The study's primary objective is to identify the most effective decision tree algorithms for spam email classification. We used the Waikato Environment for Knowledge Learning (WEKA) data mining tool in this study to implement decision tree algorithm techniques on a set of emails obtained from the University Utara Malaysia's Computer Centre. The dataset is collected at various time to ensure that the selected sample contains the required diversity. The remainder of this article will be structured as follows: The following section discusses the context for this study and related works. It is followed by a section describing the methods used to accomplish the study's objectives. Following that, a section is devoted to the study's findings, and finally, there is a conclusion section.

2 Background and Related Works

This section presents the background and related work of the study include spam filtering technique, related work on spam filtering technique, and classifications algorithms related work.

2.1 Spam Filtering Techniques

The growing number of email users has irritated the public over the rise in spam. Email spam is inconvenient for users and costs email server owners and Internet service providers a lot of money to classify. There are numerous spam filtering techniques that fall into several categories. This study will summarize it.

(i) User-defined filters: In this technique, filters automatically determine and remove spam messages based on user-defined rules. These include establishing guidelines for the acceptable subject matter and sources. For example, the user can configure his filter to reject all emails that contain a particular word in the header or emails from senders [8, 9].

(ii) Header filters: This technique relies on inspecting the headers of incoming emails for forgery indicators. Numerous spammers fabricate these headers in order to conceal their identities or locations. Normally, a header contains a wealth of information, including the recipient, subject fields, and sender; it also contains information about the servers that delivered this email, referred to as the relay chain. A good header filter can detect the forged header. Not all spammers, however, forge this information [8, 9].

(iii) Language filters: This technique handles any email that is not written in the language selected by the user. Whether or not the user uses email to communicate with anyone in a foreign language, this technique is beneficial for non-native speakers [8, 10].

(iv) Content filters: This technique is based on a set of rules that analyse the text of emails to determine whether they are spam or not. Unfortunately, this technique has some drawbacks, as these filters may exclude useful emails such as newsletters and other emails that the user specifically requests [5, 8].

(v) Blocklists and allow lists: This technique enables the user to create a blocklist of email addresses from specific users or websites and an allowed list of email addresses that are acceptable. The filter will automatically accept emails from any source on the allow list and will reject emails from any source on the block list. It takes a long time to develop this type of filter. Additionally, it is relatively easy to fool a user by using variations of well-known phrases that are still readable to the user (e.g., p*a*y) or by using bogus email addresses. As a result, new rules must be added on a continuous basis to maintain the filter's efficiency [8, 10].

(vi) Community blacklists: This technique is based on user interaction with spam filtering techniques such as black lists and whitelists. The user of this technology will automatically become a part of the millions of other users fighting spam. Numerous anti-spam techniques make use of this fact to effectively eliminate spam. If several users flag a message as spam, the technique flags it as spam and blocks it from being seen by other users [8, 10].

(vii) Bayesian analysis: Bayesian analysis is a machine learning technique that enables users' systems to learn what constitutes spam and analyses incoming messages using complex algorithms to determine whether they are spam or legitimate. This technique determines the probability that an email is spam based on previously recognized spam versus email considered acceptable. As one of a large number of users subscribed to an anti-spam technique that makes use of Bayesian filtering, the user benefits from community experience and an exact analysis system [11, 12].

2.2 Related Works

Unwanted email messages were first identified as a problem in a 1975 Internet Request for Comments [13]. Once spam or unwanted email becomes prevalent, and the emergence of this problem and the growing circle of spam-affected users. Uncontrolled and manual filtering, on the other hand, is impossible. Numerous businesses and institutions have made sustained efforts to eradicate this phenomenon; numerous techniques have been proposed, and numerous studies have been conducted on email spam filtering techniques [13]. Additionally, these organizations sought to ascertain the economic impact and disadvantages of the email service on its recipients. These efforts resulted in the discovery of numerous studies and numerous strategies, which have been implemented on a variety of datasets to achieve the desired result [3, 6].

In [14], the researcher investigated fourteen classification methods and conducted a cross-experiment to compare them, including Naive Bayesian, linear squares fit, decision tree, neural network, and Rocchio. The researcher discovered that KNN (k-nearest neighbours) is one of the best performers when scaling up to noisy and very large classification problems. A model proposed in [14] for classifying personal emails using Neural Networks and PCA (Principal Component Analysis) as a pre-processor to reduce the data's size and dimensionality.

The researchers in [15] used the minimum distance-to-mean (MDM), maximum likelihood classification (MLC), linear discrimination analysis (LDA), and neural network-based multi-layered perception (MLP) classifiers, as well as the popular C4.5 decision tree. The purpose of this study is to compare the learning ability, generalization ability, and speed of these classifiers. The results of this study demonstrated that MLP is a consistently superior classifier due to its generalization ability. In comparison, the widely used MLC classifier is not noticeably faster. On the one hand, the MDM classifier has a significant speed advantage; on the other hand, [16] compared the prediction accuracy, complexity, and training time of thirty-three classic and novel classification algorithms.

As a conclusion to this section, prior research and related work demonstrate that: there are numerous email spam filtering techniques, each with its own set of advantages and disadvantages. These techniques vary in their approach to remove or reduce spam. Some are dependent on the user, such as the community blacklists technique [8], while others operate independently of the end user, such as the Bayesian analysis technique [11]. Multiple studies indicate that using decision tree algorithms to classify email text is quite effective and efficient at detecting spam. Therefore, we set out to find the best decision tree algorithm for spam email classification.

3 Methodology

This section details the combination of methods used to accomplish the research's stated objective. The proposed methodology consists of six steps: data collection, data pre-processing, feature selection using document frequency, feature weighting using TF-IDF, J48 technique implementation with stratified cross-validation, evaluation of subject vs. body corpora, and comparison to another classification method. These steps are depicted in Fig.3, and the remainder of this section details each one.

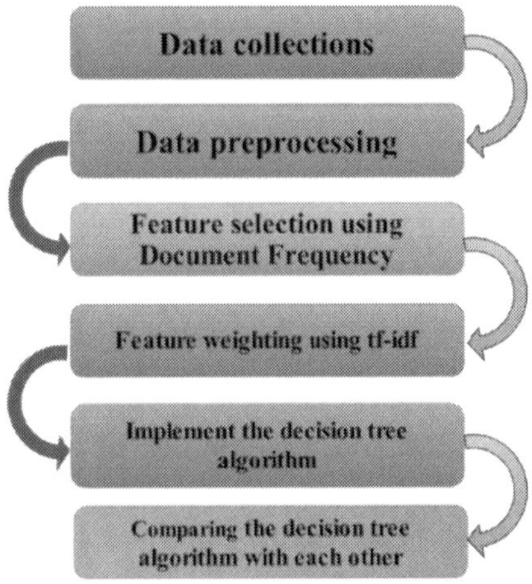

Fig. 1: Methodology Stages

3.1 Data Collections

As previously stated, the required data set was collected over a one-week period from the University Utara Malaysia's Computer Centre to achieve the required diversity in the sample of selected emails. The dataset contains approximately 100 emails, 30 of which are spam and 70 of which are legitimate.

3.2 Data Pre-processing

For each corpus, a bag-of-words representation was used (Spam and Legitimate email). The bag-of-words model is a "simplifying representation used in natural language processing and information retrieval (IR)" [17]. It represents a text (such as a sentence or a document) as a bag (multiset) of its words, disregarding grammar and even word order but retaining multiplicity.

Following that, the data is processed by extracting individual words from the corpus. This step generates a large corpus of individual words divided into two corpora (the first contains 9841 words extracted from spam email, while the second contains 11438 words extracted from legitimate email).

We then eliminate stop-words from the corpus. After removing Malay and English stop words from each corpus, the number of features decreased significantly. We extracted 2316 features from spam emails and 2824 from the body of legitimate emails. Fig. 2 illustrates the dramatic increase in the number of features prior to and following the removal of the stop-words process. The remaining features, however, are still too large to be used in the classification technique. As a result, we must reduce the number of features using a technique described in the following section.

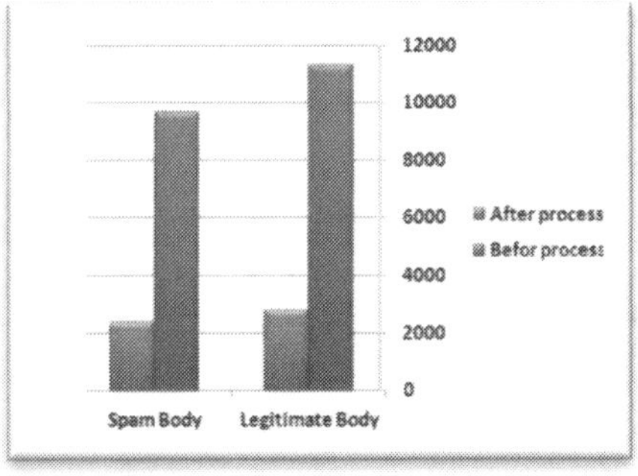

Fig. 2: Features number before and after remove stop-words process

3.3 Feature Selection

Feature selection is often an essential data processing step before applying a learning algorithm. The removal of irrelevant and redundant information often improves the performance of machine learning algorithms [5, 11, 18]. Furthermore, feature selection is an important issue in building classification models, and feature selection is advantageous in that we are thus able to limit the number of input features in a classifier to have a good predictive and less computationally in-tensive model [19].

For feature selection, we used Document Frequency (DF) method, which means for each corpus separately:

- We computed the DF score for each word, document frequency (DF) is defined as "the number of times that term occurs in the document" [20]).
- We selected the top 100 words from each corpus (the ones with the highest DF score) to balance the two corpora in feature selection. We merged the features of two corpora (spam and legitimate) after adding a classification class, which are 1 for spam and 0 for legitimate features.

3.4 Feature Weighting Using TF-IDF

Term frequency-inverse document frequency (TF-IDF), "is a numerical statistic that reflects how important a word is to a document in a collection or corpus. It is often used as a weighting factor in text mining. The TF-IDF value increases proportionally to the number of times a word appears in the document" [21]. Essentially, TF-IDF determines words' relative frequency in one document compared with the inverse proportion of the same word over the corpus of the entire document. This calculation can determine how a given word is relevant in a particular document.

3.5 Implement Decision Tree Algorithms with Stratified Cross-Validation

To evaluate the performance of the decision tree algorithms in email classification. We use cross validation, or sometimes called rotation estimation, it is an elaborate method, "if we suppose that a number of folds n is specified. The dataset is randomly reordered and then split into n folds of equal size. In each iteration, one-fold is used for testing and the other n-1 folds are used for training the classifier".

The test results are collected and averaged over all folds. This gives the cross-validation estimate of the accuracy. The folds can be purely random or slightly modified to create the same class distributions in each fold as in the complete dataset. In the latter case, the cross-validation is called stratified. For each 10-fold cross validation run, the program prints the accuracy on the test set, and at the end the average accuracy over the 10 runs. In this work, we used 10-fold cross validation to compare the performance of the classifiers on the two corpora (body and subject).

3.6 Comparing the Decision Tree Algorithms with Each other

To discover which is the best decision tree algorithms for the classification of emails between legitimate and spam, we compare the accuracy results of these algorithms with each other. In other words, we applied these techniques to UUM emails corpora. After that, we compare them by using the weighted average for precision, recall, and F-Measure scores. To achieve the goal of this study, which determines the best decision tree algorithms with the best outcome in emails classification.

4 Result and Conclusion

This section presents the comparative result of the 6 decision tree algorithms' performance: Random Forest, LMT, Decision Stump, J48, Random Tree, and REP Tree. The algorithms are compared in terms of their precision, recall, and F-measures.

4.1 Evaluation of Decision Tree Algorithms

After implementing the decision tree algorithms for the email's features corpora, we evaluated the results to choose the best one between them. The evaluation is based on three weighted average scores: precision, recall, and F-measure. Precision "is the proportion of retrieved items that are relevant, measured by the ratio of the number of relevant retrieved items to the total number of retrieved items" [17]. The recall is " the proportion of relevant items retrieved, measured by the ratio of the number of relevant retrieved items to the total number of relevant items in the collection."

F-Measure is "2 * Precision * Recall/ (Precision + Recall), a combined measure of precision and recall" [17], and the third one is recall rate. Table 1 shows the average precision, recall, and F-measure score for each decision tree algorithm. Moreover, Fig. 3 shows the dramatic difference in the weighted average between these decision tree algorithms.

Table 1. The average precision, recall, and F-measure the score for each decision tree algorithm.

	Precision	Recall	F-Measure
Random Forest	58%	41.1%	42%
LMT	11.5%	68.5%	68.5%
Decision Stump	89%	89%	89%
J48	86.5%	86.5%	86.5%
Random Tree	23%	42.5%	29.8%
REP Tree	25%	50%	33.3%

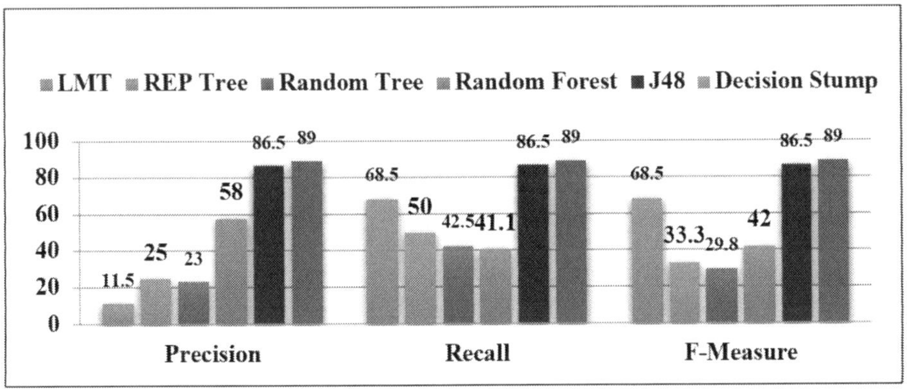

Fig. 3: The Different Accuracy Ratios Between Classifiers When They Applied on Emails Corpora.

As can be seen in figure 3, for the email's corpora, the precision, recall, and F-measure the score for the Decision Stump Algorithm is higher than the other five com-pared algorithms that are 89% for all measures. J48 comes after with 86.5% for all measures. The precision, recall, and F-measure scores for the Random Forest algorithm are 58%, 41.1 %, and 42% respectively. The precision scores for the LMT algorithm are 11.5 % and 68.5% for both recall and F-measure scores. For the REP Tree algorithm, the precision, recall, and F-measure scores are 25 %, 50%, and 33.3% respectively. And finally, the precision, recall, and F-measure scores for Random Tree are 23%, 42.5%, and 29.8% respectively.

5 Conclusion

In conclusion, after examining the performance of 6 decision tree algorithms in the classification of emails and compared its performance by implementing the algorithm on dataset corpora, we have found that the Decision Stump algorithm is the best spam emails classification algorithm for the University Utara Malaysia email service. In addition, the method of applying document frequency on the email features has reduced the number of features on the email dataset without affecting the classification accuracy. The calculation of the tf-IDF weighting technique on the reduced features proved to produce better features to be used in the classification process. This study also discovered that we had proposed the best algorithm to deal with the spam problem in email services. As for recommendations of this study, the proposed methods can also be implemented on the other email system in any part of the world due to the inclusion of language stop word removal in the pre-processing of the email data. Future work in this area includes investigating other pre-processing, feature selection, and feature weighting sections that can better represent email data to increase the accuracy of classification.

6 Acknowledgments

We want to thank the Computer Centre in University Utara Malaysia for their support in providing the data for the study.

References

[1] T. Verma, N. S. J. I. J. o. I. T. Gill, and E. Engineering, "Email Spams via Text Mining using Machine Learning Techniques," **9**, no. 4, pp. 2535-2539, (2020).
[2] N. Saidani, K. Adi, M. S. J. C. Allili, and Security, "A semantic-based classification approach for an enhanced spam detection," **94**, p. 101716, (2020).
[3] H. Taylor, "Making Mass-Spamming Illegal Rises," Harris Interactive (2011).
[4] P. Heymann, *et al.*, "Fighting spam on social web sites: A survey of approaches and future challenges," *Internet Computing, IEEE,* **11**, pp. 36-45, (2007).
[5] Clearbridge. *What is the global cost of spam?* Available: http://www.mailshine.com/2011/06/whats-the-globalcost-of-spam/(2011).
[6] M. Fossi, *et al.*, "Symantec global internet security threat report," *White Paper, Symantec Enterprise Security,* **1**, (2013).
[7] X. Guo and Z. Xia, "Fighting spam," *University of California Berkeley,* (2012).
[8] S. Youn and D. McLeod, "A comparative study for email classification," in *Advances and Innovations in Systems, Computing Sciences and Software Engineering,* ed: Springer, pp. 387-391 (2007).
[9] I. Firdausi, *et al.*, "Analysis of Machine learning Techniques Used in Behavior-Based Malware Detection," in *Advances in Computing, Control and Telecommunication Technologies (ACT), 2010 Second International Conference on,* pp. 201-203, (2010).

[10] S. T. Maller, "Email filtering methods and systems," ed: Google Patents, (2006).
[11] D. Cook, *et al.*, "Catching spam before it arrives: domain specific dynamic blacklists," in *Proceedings of the 2006 Australasian workshops on Grid computing and e-research-* **54**, pp. 193-202, (2006).
[12] P. Warkhede, *et al.*, "Fast packet classification for two-dimensional conflict-free filters," in *INFOCOM. Twentieth Annual Joint Conference of the IEEE Computer and Communications Societies. Proceedings. IEEE*, 2001, pp. 1434-1443 (2001).
[13] P. O. Boykin and V. Roychowdhury, "Personal email networks: An effective anti-spam tool," *arXiv preprint cond-mat/0402143,* (2004).
[14] A. W. Moore and D. Zuev, "Internet text classification using bayesian analysis techniques," in *ACM SIGMETRICS Performance Evaluation Review*, pp. 50-60, (2010).
[15] N. J. Kawale and S. Y. Sait, "A Review on Various Techniques for Spam Detection," in 2021 International Conference on Artificial Intelligence and Smart Systems (ICAIS), pp. 1771-1775: IEEE, (2021).
[16] O. El Kouari, H. Benaboud, and S. Lazaar, "Using machine learning to deal with
Phishing and Spam Detection: An overview," in Proceedings of the 3rd International
Conference on Networking, Information Systems & Security, pp. 1-7, (2020).
[17] M. Sahami, *et al.*, "A Bayesian approach to filtering junk e-mail," in *Learning for Text Categorization: Papers from the 2008 workshop*, pp. 98-105, (2008).
[18] B. Cui, *et al.*, "On effective e-mail classification via neural networks," in *Database and Expert Systems Applications*, pp. 85-94, (2005).
[19] G. Leroy and T. C. Rindflesch, "Using symbolic knowledge in the UMLS to disambiguate words in small datasets with a naive Bayes classifier," *Medinfo,* **11**, pp. 381-385, (2004).
[20] E. Byvatov, *et al.*, "Comparison of support vector machine and artificial neural network systems for drug/nondrug classification," *Journal of Chemical Information and Computer Sciences,* **43**, pp. 1882-1889, (2009).
[21] A. Lorenz, *et al.*, "Comparison of different neuro-fuzzy classification systems for the detection of prostate cancer in ultrasonic images," in *Ultrasonics Symposium, 2005. Proceedings., 2005 IEEE*, , pp. 1201-1204 (2005).
[22] N. Widiastuti, "Convolution neural network for text mining and natural language processing," in *IOP Conference Series: Materials Science and Engineering*, **662**, no. 5, p. 052010: IOP Publishing, (2019).

Impact Of Load Balancing Interference Routing Metric On Multi Channel Multi Radio Wireless Mesh Network

Mohammad Siraj[1*] *and Zeeshan* Abbasi[2]

[1]Department of Electrical Engineering, College of Engineering, King Saud University, Riyadh.
[2]University Polytechnic, Jamia Millia Islamia, New Delhi

Abstract. Wireless Mesh Network (WMNs) is a promising technology that has been accepted favorably by the Internet Service Providers (ISPs). ISPs use it to provide last mile broadband Internet connectivity to the end users. This popularity arises from the fact, that it inter operates well with diverse wireless systems and provides robust fault tolerance with a high degree of redundancy and reliability. Routing metrics plays a crucial part in the performance of WMN. When routing protocols are implemented, the routing metric is assigned to different paths. It calculates the optimum path to predict the best routing path. These are integrated in routing protocols to improve WMNs efficiency in terms of reliability, latency, throughput, error rate and cost. This paper compares the impact of using load balancing interference aware routing metric (LBIARM) on the performance of Multi-Channel Multi Radio WMNs (MCMR WMNs). The performance of LBIARM was evaluated by comparing it with popular standard metrics weighted cumulative expected transmission time metric and Hop Count metric. OPNET Modeler 17.5 PL1 was used as a simulation tool for implementation. Simulation results show the effectiveness of using LBIARM metric on MCMR WMNs.

1 Introduction

Wireless Mesh Networks are a promising wireless technology used for various applications like video on demand, healthcare systems, disaster and emergency, modern intelligent transport systems, public security systems, broadband internet for home and campus networking. Routing metrics is a crucial part of WMN routing protocols, which affects the performance of WMNs [1-9]. Due to the presence of static nodes and common wireless medium, it is a challenging task for researchers to design a best quality routing metric. Figure 1 shows the architecture of a WMN. IEEE 802.16 Mesh Networks comprises of two types of nodes. The two types of nodes are subscriber station (SS) and Base Station (BS). The SS is analogous to the terminal at the client end through which mobile subscribers access it via air interface. WirelessMan-OFDM is the air interface. The BS acts as an interface between the external and IEEE 802.16 Network.

*Corresponding author: siraj@ksu.edu.sa.

2 Related Work

Researchers have designed many routing metrics. Some of the popular are Expected Transmission Count (ETX) Metric [10], Expected Transmission Time (ETT) Metric[11], Weighted Cumulative Expected Transmission Time (WCETT) Metric [11], Load Aware Expected Transmission Time (LAETT) [14], Metric of Interference and Channel Switching (MIC)[12-13], Exclusive Expected Transmission Time (EETT)[15], Interference Load Aware (ILA) metric [16], Interference Aware Routing Metric (iAWARE)[17], Load Balancing Interference Aware Routing Metric (LBIARM) [18], Expected Dynamic Transmission Cost (EDTT) [19] and Physical layer Metric[20] for Internet of things[21] and Opportunistic routing metrics[22].
To evaluate the effect of routing metric on the performance of WMNs, we consider our metric with two popular metrics for our performance evaluation. These three metrics are briefly described below:

2.1 Hop Count Metric

The most common metric used in the multi-hop routing protocols is Hop-Count metric [10]. It is used in protocols like DSDV, DSR and AODV. Hop-Count metric finds route having the minimum number of hops. Hop-Count can outclass other metrics, which are dependent on load in high agility situations. The metric is very stable and has the isotonicity characteristic. As a result, least weight paths can be discovered proficiently. The weakness of this metric is that it does not address interference, channel diversity, varying load of the link and capacity of the load. It finds path having poor throughput and high packet loss ratio. This is because the links that are slower, take a lot of time to transmit packets.

2.2 Weighted Cumulative Expected Transmission Time (WCETT) Metric

The most WCETT was presented by Draves [16] to enhance the ETT metric in the multi radio mesh networks by taking into account the diversity of the channels. The WCETT metric of a path p is defined as follows:

$$WCETT_p = (1-\alpha) * \sum ETT + \alpha * MaxX_j \quad (1)$$

Where X_j = summation of links ETT values on channel j

$$X_j = \sum_{\text{hops on channel j}}^{n} ETT_i \quad 1 \le j \le k \quad (2)$$

Where,
α is a tunable parameter $0 \le \alpha \le 1$ which controls the preferences over the path length versus channel diversity.

WCETT is composed of weighted average of two components. The first term sums up individual link ETTs while the second term sums up the ETTs of every link of a given channel. This adds channel diversity to the routing metric causing low intra-flow interference. Using WCETT, multi radio wireless mesh network's performance is enhanced in comparison to ETX, ETT and Hop-Count metrics. WCETT metric is not isotonic and due to this, it can't be used with link state routing protocols. Secondly, the inter-flow interference effects are not explicitly taken into account by WCETT. As a result of this, sometimes paths are created, which have high levels of interference.

 https://doi.org/10.1051/itmconf/20224201002

2.3 Load Balancing Interference Aware Routing Metric (LBIARM)

LBIARM [20] is an enhancement of WCETT comprising of two components. The first component is same as WCETT component. The second part considers channel diversity.

$$LBIARM = (1-\alpha)\sum_{i\in p} ETT_i + \alpha \sum_{i\in p} ETT_i * N_i \tag{3}$$

Where N_i is the set of interfering links on link I and.

$$ETT = ETX * \frac{S}{B} \tag{4}$$

Where,
S = average packet size.
B = Bandwidth of the existing link.
α is a tunable parameter 0 ≤ α ≤ 1.

3 Performance Evaluation

The simulation setup to validate and evaluate various routing metrics, including LBIARM, Hop Count and WCETT in OPNET Modeler is discussed here. The three routing metrics namely Hop Count, WCETT and LBIARM are compared for a 4 X 4 grid mesh network as shown in Figure 1.

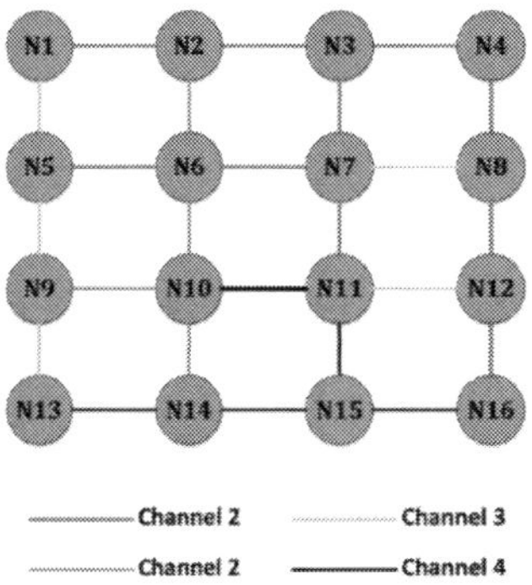

Fig. 1. 4x4 Grid Wireless Mesh Network with 4 channel assignment

In the simulated wireless mesh network, 16 static mesh nodes are deployed in a grid scenario in 800 x 800 m2 area. Identical distance is maintained between each one hop node pair. Similarly, the interference range is kept almost equivalent as all the mesh routers have same transmission strength. One of the mesh routers, which is centrally located in the network, was selected to act as a gateway node. Each of the mesh nodes has multi radio that was allocated to different channels. The same channel allocation scheme was extended for all the mesh routers as in [4]. That is, each mesh router had more than one interfaces, and these interfaces were allocated to different channels.

The source nodes send Constant Bit rate (CBR) traffic with UDP as the transport protocol. It consists of 500, 1000 and 1500 byte packets with a transmission rate of 20 packets/second. OPNET source code was modified to calculate LBIARM at each node. Interference traffic load was created by broadcasting HELLO messages at an interval of 1 second periodically to all neighbor nodes on channel i. Upon receiving this message, the neighbor nodes update the traffic load information of the corresponding nodes in their neighbor's table. Simultaneously, information of average packet buffered at the nodes is updated. For calculation of LBIARM, HELLO messages are also used. The value of parameter was taken as 0.3 for calculation of WCETT and LBIARM. This value was taken, as it was found that optimum value of WCETT is when α =0.3. These simulations were executed on a Pentium XEON X5560 2.79 GHz 48 GB RAM. The simulation parameters are illustrated in Table 1.

Table 1. Simulation Parameters.

Parameter	**Value**
Network Scenario	Campus Network
Network Grid	800 X 800
Number of Nodes	16
umber of radios	1,2 and 4
Number of Channels	4
Packet Size	500,1000 and 1500 Bytes
Interference range	400 m
Traffic Model	Constant Bit Rate(CBR)
Transmission Power	10 mW
Queue size at Routers	50 Kbytes
Physical layer protocol	PHY 802.11g
CBR sender's rate	20 packets/sec
Tx. rate at Physical layer	54 Mbits/sec

The routing metrics are compared based on the following performance metrics.

- **Packet Delivery Ratio (PDR):** It is defined the ratio of total number of data packets that are received by the destination node to the total number of data packets transmitted by the source node.
- **End-to-end delay of data packets:** It is the time taken by the data packet to reach from source node to destination node.

Figure 2, shows the throughput comparison with multi radios. Initially, throughput was observed with the number of radios taken as 1. Then the number of radios were increased to 4. A marginal difference in throughput was observed between WCETT and LBIARM when single radio was chosen. When the number of radios was increased to 4, a substantial difference was observed between LBIARM and WCETT. This is because there is a heavy contention for the physical medium when multiple flows are directed towards single radio. Contention occurs at the link layer, which is the reason of delay at each hop. As the number of hops are increased, interferences increases leading to more contention due to the receive queue, which is very high and there is a buffer overflow. When more than one radio is used the time slots to send the packets also reduces and the minimal link satisfaction ratio increases, as a result network performance shows enhancement. It is observed that LBIARM selects higher performance paths. LBIARM achieves a throughput enhancement of 80% and 16% respectively over Hop Count and WCETT respectively for four radio scenarios.

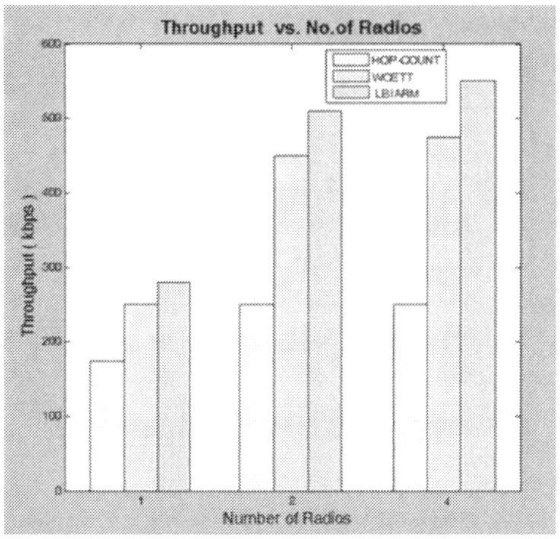

Fig 2. Throughput versus Radios

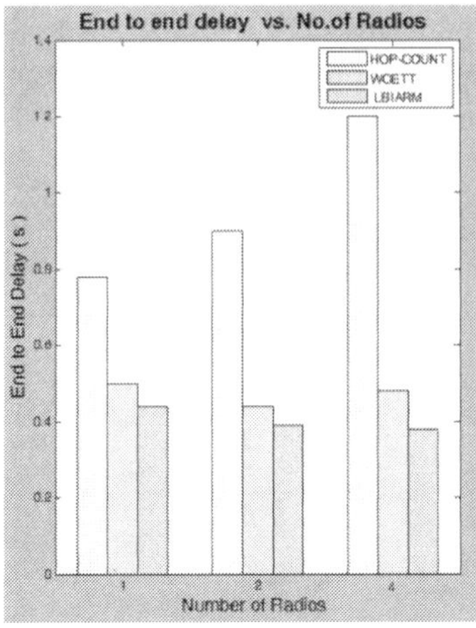

Fig 3. End to end delay versus radios

Figure 3, shows the comparison between an end-to-end delay versus multi radios. It is observed that as the numbers of radios are increased, end to end delay in Hop Count is increased whereas even after increasing number of radios, there is not much change in WCETT. This is because Hop Count has not been designed for multi radio. There is not much improvement in WCETT, as it does not recognize inter-flow interference. So, if inter-flow interference exists, the packets are routed towards gateway leading to a large queue. LBIARM has a shorter delay in comparison to others as it selects routes whose load is less than the channel capacity, thereby minimizing the delay time in queuing, collision and buffer overflow. When more than one radio is selected, it is observed that LBIARM can improve end to end delay by minimizing the latency by 80% and 16 % from Hop Count and WCETT respectively.

Figure 4, shows performance comparison between throughput and the packet size when packet size is varied from 500 to 1500. The simulation result shows that WCETT and LBIARM perform well when the packet becomes large. The improvement is around 80 % and 13 % from Hop Count and WCETT respectively.

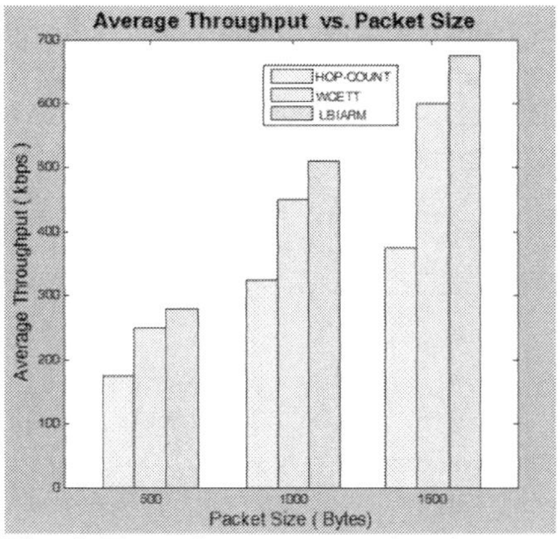

Fig. 4. Throughput versus packet size.

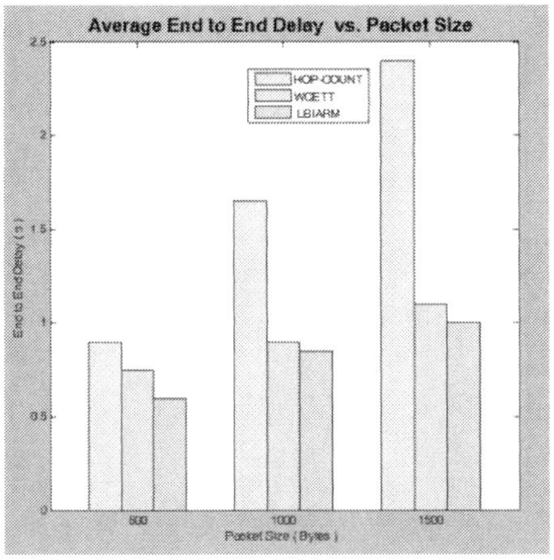

Fig. 5. End to end delay versus packet size.

Figure 5, shows performance comparison between end-to-end latency and the packet size when packet size is varied from 500 to 1500. LBIARM considers routing overhead in packet transmission. When data packet size is small, the overhead takes a higher percentage of the transmission time compared with large data size. Therefore, LBIARM performs much better with smaller packet size as compared to the larger packet size. There is an improvement in end-to-end delay as compared to Hop Count and WCETT. Delay improvement by for a large packet nearly by 71 % and 10 % from Hop Count and WCETT respectively for a four-radio scenario.

4 Conclusion and Future work

In this work, it is seen that the use of a load balancing interference aware routing metric enhances the performance of the network by minimizing the end-to-end delay, thereby increasing the throughput of the network. It is observed that the usage of a normal metric results in a larger end to end delay and a decreased throughput because of intra-flow interference, inter-flow interference and traffic load in the network. The load balancing interference aware routing metric takes into account intra-flow interference, inter-flow interference and traffic load. These parameters are very critical when the number of nodes increase. It will be very interesting to observe, how load balancing interference aware routing metric performs in WMN test bed.

References

1. M.Siraj and K.A. Bakar, "Link establishment and performance evaluation in IEEE 802.16 wireless mesh networks,". Int. J. Phys., vol.6 issue 136, pp.3189-3197, (2011)
2. M. Siraj "A Survey on routing algorithms and routing metrics for Wireless Mesh Networks", World Applied Sciences Journal, vol.30 issue 7, pp.870-886, (2014).
3. M. Siraj and K.A. Bakar, "Minimizing Interference in Wireless Mesh Networks Based Telemedicine System", Journal of Computer Science, Vol.8. Issue8, pp.1263-1271, (2012).
4. M. Siraj and K.A. Bakar, "To Minimize Interference in Multi Hop Wireless Mesh networks using load balancing interference aware protocol", World Applied Sciences Journal 18(9):1271-1278, vol.11 No. 3, pp 1652-1664, (2013).
5. M. Siraj, M. Shoaib and S. Alshebeili, "Performance enhancement of cognitive radio Wireless Mesh Networks by link scheduling and beamforming," Saudi International Electronics, Communications and Photonics Conference, pp. 1-6, (2013).
6. M. Siraj and Z. A. Abbasi, "An Efficient Video on Demand System over Cognitive Radio Wireless Mesh Networks," 2013 International Symposium on Computational and Business Intelligence, pp. 231-234, (2013).
7. M. Siraj and S. Alshebeili, "RPCRAN: A Routing Protocol for Cognitive Radio Ad hoc Networks", International Journal of Innovation and control Volume 9, Number 10, (2013).
8. M. Siraj and K.A. Bakar, "To Minimize Interference in Multi-Hop Wireless Mesh Networks Using Load Balancing Interference Aware Protocol", World Applied Sciences Journal, Vol.18. Issue9, pp.1263-1271, (2012).
9. Mohammad Siraj and Saleh Alshebeili "Performance Enhancement in multi hop Cognitive Radio Wireless Mesh Networks", International Journal of Innovation and control, Volume 9, Number 10, (2013).

10. De Couto, D. Aguayo, J. Bicket and R. Morris, "A High-Throughput Path Metric for Multi-Hop Wireless Routing," in Proc. of the 9th annual international conference on Mobile computing and networking, pp 134 – 146, (2003).
11. R.Draves, J. Padhye and B. Zill. "Routing in multi radio, multi hop wireless mesh networks," In proceedings of ACM MOBICOM.pp.114-128, (2004)
12. Y. Yang, J. Wang. and R. Kravets, "Designing routing metrics for mesh networks, Proc. WiMesh," (2005)
13. Y. Yang, J. Wang. and R. Kravets.," Interference-aware loop-free routing for mesh networks," (2006).
14. H.Aiache, V. Conan, L. Lebrun and S. Rousseau, "A load dependent metric for balancing Internet traffic in Wireless Mesh Networks," Mobile Ad Hoc and Sensor Systems, pp.629-634, (2008).
15. W. Jiang, S. Liu, Y. Zhu and Z. Zhang, "Optimizing Routing Metrics for Large-Scale Multi-Radio Mesh Networks" Wireless Communications, Networking and Mobile Computing, pp.1550-1553, (2007).
16. D.M. Shila and T. Anjali, "Load-aware Traffic Engineering for Mesh Networks. Computer Communications and Networks," pp.1040-1045, 2007
17. A.P. Subramanian, H. Gupta, S. Das, and J. Cao (2008). "Minimum Interference Channel Assignment in Multi-radio Wireless Mesh Networks," IEEE Transactions on Mobile Computing, vol.7 issue 12, pp.1459-1473, (2008).
18. M. Siraj and K.A. Bakar, "A Load balancing Interference Aware Routing Metric (LBIARM) for multi hop wireless Mesh Network,", Int. J. Phys. Sci., vol.7 issue 3, pp 456-461, (2012).
19. Cheng, H., Wang, C., Zhang, X., "An opportunistic routing in energy-harvesting wireless sensor networks with dynamic transmission power". IEEE Access 7, 180652–180660p, (2019)
20. M. Abdullahi, K. Malekinasab, W. Tu and M. Bag-Mohammadi, "An Efficient Metric for Physical-layer Jammer Detection in Internet of Things Networks," IEEE 46th Conference on Local Computer Networks (LCN), pp. 209-216, (2021).
21. Islam, N.; Altamimi, M.; Haseeb, K.; Siraj, M. Secure and Sustainable Predictive Framework for IoT-Based Multimedia Services Using Machine Learning. Sustainability,13, 13128, (2021)
22. Mostafa Abdollahi, Farshad Eshghi, Manoochehr Kelarestaghi, Mozafar Bag-Mohammadi,Opportunistic routing metrics: A timely one-stop tutorial survey, Journal of Network and Computer Applications, Volume 171,102802, ISSN 1084-8045, (2020)

Face Recognition System Using Independent Components Analysis and Support Vector Neural Network Classifier

Mustafa A. Shihab[1*], *Mustafa* Zuhaer Nayef Al-Dabagh[2*], *Ali* jawad kadhim Alrubaie[3], *Murthad* Al-Yoonus[4], and *Mohammed* Ghazal[5]

[1]Department of Computer Science, College of Science, Tikrit University, Tikrit, Iraq
[2]Department of Computer Science, College of Science, Knowledge University, Erbil, Kurdistan Region, Iraq
[3]Research and Studies Unit, Al-Mustaqbal University College, 51001 Hillah, Babil, Iraq
[4]Computer Science Department, Engineering and Science College, Bayan University, Kurdistan Region, Iraq
[5]Computer Technology Engineering Department, Technical Engineering College of Mosul, Northern Technical University, Mosul, Iraq

Abstract. With an increasing number of security threats in recent years, the field of automatic facial recognition has seen many new developments. The introduction of many new face recognition algorithms focuses on increasing the accuracy rate of the recognition system. This paper introduces a face recognition system using Independent Component Analysis (ICA) for feature extraction and a Support Vector Neural Network (SVNN) for classification. As well as introducing a comparison between SVNN and Support Vector Machine (SVM) and Artificial Neural Network (ANN) classifiers, they are applied to prove the reliability of the proposed method. The implemented experiments use Yale databases, and the results prove that the proposed approach has a higher recognition rate than the (ICA+SVM) and (ICA+ANN) approaches for face recognition.

1 Introduction

Facial recognition systems use computer-coded algorithms to spontaneously recognize or verify a person using digital images or video frames [1]. Many methods compare selected facial image features with a facial database [2]. Typically, this technology pertains to security systems and is analogous to other biometrics solutions, such as fingerprints or eye-iris recognition systems [3].

The PCA method [4] is unquestionably advantageous for feature extraction, but it is better suited for image reconstruction because it does not take into account the various classes of separability. In contrast, the LDA method (Linear Discriminate Analysis) is better than the PCA because it addresses the feature subspace deficiencies of the PCA in terms of separability [5].

* Corresponding author: mustafa.zuhaer.nayef@gmail.com

On the other hand, the ICA method provides an improved basis for recognizing high-order relationships between the image's pixels [6]. Most recently, popular machine learning methods use large margin classifiers, for instance, SVM [7]. SVM pattern classification even applies within an ample representation space. Establishing the hyperplanes for pattern classification with that method is imperative. The derivation of the hyperplanes is a ratio of the number of classes. The identification of unseen class elements can be accomplished by calculating posterior probabilities in 1-vs-1/1-vs-rest binary SVMs (Piatt's method) or multi-class SVMs [8].
Training of an ANN uses patterns from the classification process to ascertain the weight collection. ANN is implementing a threshold classification method to define an outside pattern. If the ANN ascertains that the patterns do not apply in any training of an ANN, it uses patterns from the classification process to ascertain the weight collection. ANN is implementing a threshold classification method to define an outside pattern. If the ANN ascertains that the patterns do not apply in any given class, it will report the results of outside classes [9].
The organization of the contents of this paper is as follows: Section 2 introduces the preprocessing, feature extraction, and classification stages, while Section 3 outlines the experiments and analysis of the results. Finally, the conclusions summarise the paper and future work in section 4.

2 The structure of the proposed system

The structure of the proposed system consists of three significant steps: preprocessing, extracting facial features, and classification. Fig. 1 displays the structure of the proposed recognition system in which the ICA method is used to get features from the image, and a new training method named SVNN [9] is applied for classification. Finally, several experiments are sure to be compared with other training methods. The elaboration of these steps is as follows:

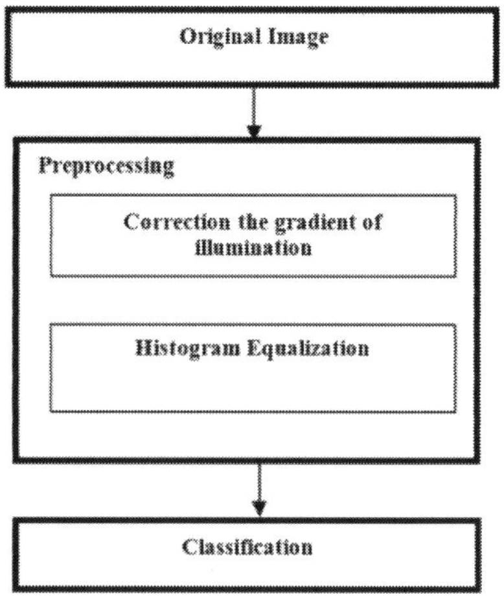

Fig. 1. Face Recognition System Process.

2.1 Image preprocessing

The image preprocessing step aims to reduce the light on the images; therefore, is necessary to implement some preprocessing steps before feature extraction. The preprocessing stage involves two steps:

2.1.1 Correction the gradient of illumination

This step calculates the best-fit brightness value in the image and subtracts this value from the original image's pixels. By using this step, the heavy shadows from extreme lighting angles can be reduced [10].

2.1.2 Histogram equalization

It is utilized to flatten the image's histogram to counterbalance the changes in illumination brightness and the differences in the camera response curves [11].

2.2 Feature Extraction

This stage aims to extract characteristic features from the image in order to employ them in the classification stage. Without applying this stage, the recognition process becomes complicated and impossible because the original face image has a tremendous amount of information, and most of this information is unusable for classification. This stage minimizes the amount of information needed to represent the facial image.

2.2.1 Independent component analysis (ICA)

ICA [12] is one of the unsupervised learning rules derived from the principle of the optimal sigmoid neuron. Let us consider that x is the signal input, y is the output, and g is a nonlinear squashing function.

$$u=wx+w_0 \tag{1}$$

$$y=g(u)=\frac{1}{1+e^{-u}} \tag{2}$$

This input x is processed by the function g(x). The output density of data depends on three components. The first is the mean, the second is the variance of fx (x) to the slope, and the third is the threshold of a nonlinear squashing function g(x). To maximize information transfer requires deriving the optimal weight (w) on (x), which best agrees with the probability density of x to the nonlinearity gradient; optimal w presents the flattest output density.
The ascending gradient of entropy of the output y concerning weight w derives the optimal weight. Multiple inputs concerning outputs necessitate maximizing the joint entropy of the output, which encourages forward statistical independence of the individual outputs. When the cumulative density functions of the underlying independent components equal the nonlinear transfer function g (up to scaling and translation), an increase in the correlative information between the input x and the output y also reduces the correlative information between the two. Where equation (3) gives the update rule for the weight matrix (W) for multiple inputs:

$$\Delta W=\left(I+y'u^T\right)W \tag{3}$$

$$\text{where } y'=\frac{\partial}{\partial y_i}\frac{\partial y_i}{\partial u_i}=\frac{\partial}{\partial u_i}\ln\frac{\partial y_i}{\partial u_i}$$

The logistic transfer function is employed:

$$g(u)=\frac{1}{1+e^{-u}}\ \text{, given } y' = (1-2y_i) \quad (4)$$

The algorithm incorporates a "sphering" action before learning [12]. Then, by subtracting the means of the row, the dataset X is derived. Then a zero-phase whitening (Wz) filter is applied to X, which is the inverse square root of the covariance matrix doubled:

$$W_z=2*(XX^T)^{-\frac{1}{2}} \quad (5)$$

This method can eliminate the first and second statistical order of the data. Then, by setting the mean and covariance to zero, they equalize the variances. Evaluating the complete transformation from the zero-mean input provides the product of the sphering matrix and the learned matrix, W1 = W * Wz. The ICA algorithm's pre-whitening filter has the Mexican-hat shape of retinal ganglion cell receptive fields; that excludes much of the variability caused by illumination.

2.3 Classification

This step aims to classify the image depending on the features extracted from every image. Then SVNN is used to classify. This method provides a maximal-margin training method and aims to maximize the margin by minimizing the values related to the parameters of the classifier model; this way reduces the complexity of the classifier without affecting the accuracy of the training data.

2.3.1 SVNN

The main idea of the SVNN [9] is not based just on the maximal-margin principle but also avoids the nonlinear SVM kernels [7]. The given SVM decision function can be shown in equation (6):

$$c(x)= \text{sgn}\left(\sum_{i=1}^{N_W} y_i\alpha_i k(x_i,x)+b\right) \quad (6)$$

Where αi and b are SVM parameters, (xi, yi) is the ith support vector data pair, the values of the (sgn) function are one of the argument is more significant than zero and minus one if it is less than zero, and K is a nonlinear kernel function. Using the equation above, the classification needs a long time when the number of support vectors is large. Using SVNN overcomes this problem. To better understand the method, it is convenient to consider the soft margin SVM optimization problem, as follows:

$$\min_{w,\varepsilon_i} = \left[\frac{1}{2}\ /|w|/^2 + C\sum_{i=1}^{N}\varepsilon_i\right] \quad (7)$$

subject to

$$\forall i\ |y_i(w\cdot x_i-b) \geq 1-\varepsilon_i \quad (8)$$

$$\forall i\ |\varepsilon_i > 0 \quad (9)$$

Separating hyperplane com poses w and b, where C is a constant, yi is the target class of the ith training case, and ξi are slack variables, which equate to the degree of misclassification the vector xi. The optimization is a trade-off between a high margin (min ||w||2), and a tiny error penalty (min C $\sum_{i=1}^{N}$ξi).). This approach is trained the NN by solving a similar optimization problem:

$$\min_{w,\varepsilon_i} = \left[\lambda_{min} + \lambda_{max} + \frac{C_1}{N}\sum_{i=1}^{N}\varepsilon_I\right] \quad (10)$$

Subject to

$$\forall i \mid y_i\hat{y} \geq 1 - \varepsilon_i \quad (11)$$

$$\forall i \mid \varepsilon_i > 0 \quad (12)$$

Where y is given by y =, W-2-T., y-h.+, b-2., and C is a regularisation hyperparameter, yi is the objective class of the ith preparation example, and i and are slack variables, as alluded to before, which ascertain the degree of classification error vector xi. By replacing the constrained optimization problem equations (10) and (12) within the equivalent unconstrained optimization problem (13), the objective function ϕ becomes discontinuous, rendering gradient-based optimization methods ineffective. As a result, a real-coded GA (Genetic Algorithm) is implemented to determine (13), utilizing as a fitness function ϕ and H(t) = max (0, 1 t) is the Hinge loss.

$$\min_{w_1,w_2,b_1,b_2} \Phi \quad (13)$$

Where

$$\Phi = \lambda_{min} + \lambda_{max} + \frac{C_1}{N}\sum_{I=1}^{N} H(y_i\hat{y}_i) \quad (14)$$

3 Experimental Results and Analysis

Experiments using three methods are implemented on the Yale database [13] to evaluate the system's performance: ICA plus SVNN, ICA with ANN, and ICA with SVM. The Yale database consists of 165 images, with different images under different lights, expressions, and perspectives for each person. The resolution of the images is 243 x 320 pixels with 265 grey levels per pixel.

The ICA with ANN is applied in the first experiment using ICA as a feature extraction method and the ANN as a classifier. The accuracy rate obtained using these two methods is 93.3333. While the second experiment uses ICA with SVM, the accuracy rate of enhancement compares favourably to the previous experiment and reaches 94.2857.

Finally, implementing ICA with SVNN is implemented. ICA is used to obtain the essential features that can be used in the next stage (classification), while SVNN is used to classify these features and overcome the significant amount of time needed by a traditional SVM method to classify a large number of support vectors. This method enhanced the results by 0.9523, which compares favourably to traditional SVM and ANN. Where the accuracy rate obtained by this method reaches 95.2380, table (I) displays the obtained results:

Table 1. **The experiments result on Yale database.**

Method	No. of training Sample	No. of testing Sample	Accuracy Rate
ICA+ANN	60	105	93.3333
ICA+SVM	60	105	94.2857
ICA+SVNN	60	105	95.2380

4 Conclusion and the Future Work

This paper proposes the combination of ICA and SVNN algorithms as an optimal model for a face recognition system. This is done by comparing three types of classifiers: ANN, SVM, and SVNN. The ICA method is an accepted feature extraction method widely used in image processing applications. The new training method called SVNN is introduced as a classifier and improves the other methods. This method is applied to increase the classification margin and decrease the time needed by the SVM method when classifying a large number of support vectors. To evaluate the performance requires the implementation of several experiments on the Yale Face database. The results then improve when applying the ICA with SVNN method, resulting in a more than 95% recognition score. Future work could use powerful ICA modifications, such as Kernel ICA. Moreover, other methods could be combined, such as K-nearest neighbour, to improve the classification.

References

1. P. M. Kumar, U. Gandhi, R. Varatharajan, G. Manogaran, R. Jidhesh, T. Vadivel, *Intelligent face recognition and navigation system using neural learning for smart security in Internet of Things*, Cluster Comput., 22, 4 (2019)
2. M. Z. N. Al-Dabagh, M. I. Ahmad, M. N. M. Isa, S. A. Anwar, *Face Recognition System Based on Fusion Features of Local Methods Using CCA*, in 2020 8th International Electrical Engineering Congress (iEECON), Chiang Mai, Thailand (2020)
3. I. Q. Mundial, M. S. U. Hassan, M. I. Tiwana, W. S. Qureshi, E. Alanazi, *Towards facial recognition problem in COVID-19 pandemic*, in 2020 4th International Conference on Electrical, Telecommunication and Computer Engineering (ELTICOM), Medan, Indonesia (2020)
4. B. M. S. Hasan, A. M. Abdulazeez, *A Review of Principal Component Analysis Algorithm for Dimensionality Reduction*, J. Soft Comput. Data Min., 2, 1 (2021)
5. S. V. Kumar, T. V Rajinikanth, S. V. Raju, *Heart Attack Classification Using SVM with LDA and PCA Linear Transformation Techniques*, Mach. Learn. Technol. Appl. Proc. ICACECS 2020, (2021).
6. V. Vapnik, *Statistical Learning Theory*, John Wiley, Sons, New York, 1998.
7. H. Wang, Y. Shao, S. Zhou, C. Zhang, N. Xiu, *Support Vector Machine Classifier via L0/1 Soft-Margin Los*s, IEEE Trans. Pattern Anal. Mach. Intell. (2021)
8. H. Tran, T. Le, N. Thuy, *Facial Expression Classification Using Artificial Neural Network and K-Nearest Neighbor*, Int. J. Inf. Technol. Comput. Sci., 7 (2015)
9. O. Ludwig, U. Nunes, R. Araujo, *Eigenvalue decay: A new method for neural network regularization*, Neurocomputing, 124 (2014)
10. S. Luo, H. Shen, H. Li, Y. Chen, *Shadow removal based on separated illumination correction for urban aerial remote sensing images*, Signal Processing, 165 (2019)
11. J. Raghavan,M. Ahmadi, *Preprocessing Techniques to Improve CNN based Face Recognition System*, in CS & IT Conference Proceedings, Kanton Zurich, Switzerland (2021)
12. A. I. Abdullah, M. Z. N. Al-Dabagh, M. H. ALHABIB, *Independent Component Analysis and Support Vector Neural Network for Face Recognition*, Int. J. Appl. Eng. Res., 13, 7 (2018)
13. P. N. Belhumeur, J. P. Hespanha, D. J. Kriegman, *Eigenfaces vs. fisherfaces: Recognition using class specific linear projection*, IEEE Trans. Pattern Anal. Mach. Intell., 19, 7 (1997)

Analysis of Encryption Algorithms Proposed for Data Security in 4G and 5G Generations

Khalid Fadhil Jasim [1*], *Kayhan Zrar* Ghafoor[2,3], *and Halgurd S.* Maghdid [4]

[1] Department of Computer Science, Cihan University-Erbil, Kurdistan Region, Iraq
E-mail: khalid.jassim@cihanuniversity.edu.iq
[2] Department of Computer Science, Knowledge University, Erbil 44001, Iraq, 3 Department of Software Engineering & Informatics, Salahaddin University, Erbil, Iraq
[3] Department of Computer Science, Knowledge University, Erbil 44001, Iraq, Department of Software Engineering & Informatics, Salahaddin University, Erbil, Iraq
[4] Department of Software Engineering, Faculty of Engineering, Koya University, Kurdistan Region-F. R. Iraq

Abstract. This paper intended to investigate and analyze encryption algorithms that can be proposed to perform data security in 4G and 5G networks. The research explores different standards, services, and features presented via 4G and 5G networks. Also, the basic components of encryption algorithms (e.g., ZUC, SNOW 3G, and AES) are investigated. For instance, initialization keys have been identified and analyzed due to their vital roles in determining the security of the encryption algorithms. Moreover, the researchers analyzed the effective elements of these algorithms (i.e., LFSR registers, substitution boxes, NLF functions like finite state machines, Math transformations, secret encryption keys, and non-secret IV keys). Cryptanalysis methods play important roles in determining the security of these algorithms. Thus, some cryptanalysis methods have been explored and investigated. Various weak points have been identified in initialization process of these algorithms. Therefore, different recommendations are presented that enhance the security of these ciphers, and can be reflected in data security in 4G and 5G networks.

1 Introduction

In the previous years, Mobile Communication Networks had various positive impacts on our life and society. Today, most people rely on Mobile devices to perform many activities such as mobile tickets, phone calls, bank transactions, internet access, and parking payment. The new generation of 5G networks supports critical communications, enhanced mobile broadband (eMBB), and used in Internet of Things (IoT) applications [1]. Also, 5G networks offer various services (e.g., Unmanned Aerial Vehicles, communication between cars, data networks using blockchain, and smart networks) [2]. More features are supported by these

* Corresponding author: khalid.jassim@cihanuniversity.edu.iq

networks like audio and video with high clarity, Artificial Intelligence abilities, and high-speed data transmissions. Furthermore, the frequency bandwidth (between 1.8 and 2.6 GHz) and support transmission speeds between (10 and 50 Gbps) [3].

The 4G networks supported mobile broadband internet accessibility. Smart devices and laptops can easily access the Internet networks. 4G networks adopted various standards such as Long-Term Evolution (LTE, as standard technology) and Worldwide Interoperability for Microwave Access (WiMAX) [4]. The 4G-LTE networks supported frequency bandwidth between (1.4 & 20 MHz). In uplink data transmissions, the bit rate (50 Mbps) and the bit rate between (100 & 326.4 Mbps) in downlink data transmissions. The 4G networks with WiMAX standards offered suitable bandwidth (1.25 … 20 MHz), and the bit rate was (40 Mbps, via the fixed stations) [5]. In terms of data security of 4G and 5G networks, various encryption algorithms are proposed to secure data transmissions via smart devices and mobile network base stations. For instance, the proposed encryption algorithms included ZUC cipher, AES cipher, and SNOW 3G cipher [6].

The research is organized as follows. Some standards and features offered in 4G and 5G networks introduced in Section 1. Section 2 provides analysis the components of AES cipher algorithm. Section 3 investigates the elements of SNOW 3G cipher. Section 4 is devoted to analyze the cryptographic elements of ZUC cipher. Section 5 covers the discussion of 4G networks, 5G networks, and the analysis results of the ciphers algorithms (i.e., ZUC, AES, and SNOW 3G). Finally, the conclusions are presented via section 6.

2 Analysis of AES Cipher Algorithm

Advanced Encryption Standard (AES) proposed according to the National Institute of Standards and Technology [7]. The AES-128 cipher contains ten rounds functions, AES-192 with 12 rounds functions, and AES-256 with 14 rounds functions. Each round consists of four mathematical transformations (SubBytes operation, ShiftRows operation, MixColumns, and AddRoundKey operation). The AES-128 cipher algorithm, which operates with counter mode technique (AES-CTR) was proposed as a secure cipher algorithm. This version used in confidentiality and integrity algorithms (EEA2 and EIA2) in 4G and 5G networks [8].In light of cryptanalysis of AES cipher algorithm, the Impossible Differential technique used to analyze the AES-128 and was conducted in [9, 10]. Boura et al. [11] introduced the Improved Impossible Differential technique, and implemented on AES cipher algorithm. This method used to analyze the AES-128 cipher (with 7 from 10 total rounds). The data complexity equal to (2^{105}), time complexity equal to ($2^{106.88}$), and memory complexity equal to (2^{74}).

The initialization phase of AES-CTR cipher algorithm depends on secret encryption key (K, K=128 bits) and counter block (T, T=128 bits) [12]. AES-CTR includes 10 rounds and 4 math transformations (SubBytes, ShiftRows, MixColumns, and AddRoundKey) [13]. In light of the Brute force attack, this cipher algorithm depends on a secret encryption key (K = 128 bits). Thus, Brute force attack requires (2^{128}). It is recommended to increase the lengths of secret key and counter block T, and it is more secure to use encrypted counter block.

3 Analysis of SNOW 3G Cipher Algorithm

The first version of SNOW 3G cipher adopted for data security in (3G-UMTS) [14]. This cipher used in some data security algorithms (e.g., 128-EEA1 and 128-EIA1) as part of 4G-LTE [15]. Moreover, the SNOW 3G cipher algorithm proposed for data security in 5G network [6]. In this context, SNOW 3G requires secret key (K=128 bits), and IV key (IV=128 bits). The first part of SNOW 3G contains LFSR (S0, S1, … S15). The second part is the finite state machine (FSM). FSM contains Registers (R#1, R#2, and R#3), 2 Substitution Boxes (Sbox(S#1) & Sbox(S#2)) (Figure 1) [16].

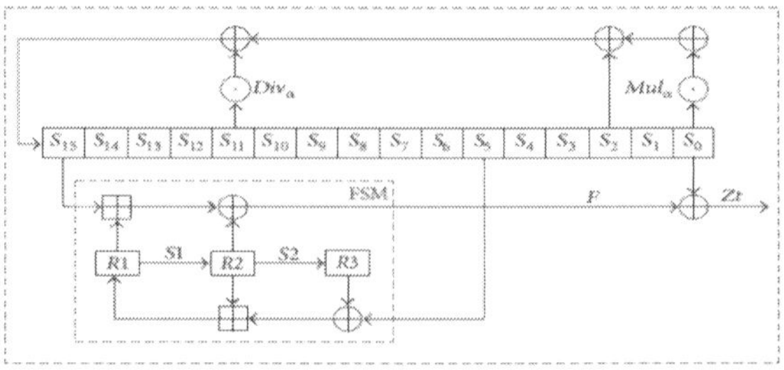

Fig. 1. Design of SNOW 3G Cipher Algorithm

In the cryptanalysis field, the Chosen IV analysis method was conducted to analyze the algorithm of SNOW 3G cipher. The distinguishing attack method applied on (13) steps of SNOW 3G, it requires data complexity (2^{8}) and time complexity (2^{8}). In case of applying choosing IV method on (14) steps of SNOW 3G, it needs data complexity ($2^{12.1}$) and time complexity (2^{27}) [17]. Also, the vectorized linear approximations used to attack the SNOW 3G. It requires (24 bits) approximation and complexity (2^{172}) [18]. Furthermore, in the initialization phase of SNOW 3G, register (LFSR) loaded with IV key and secret key (K). The key (K= K#0, K#1, K#2, and K#3), IV key (IV=IV#0, IV#1, IV#2, and IV#3). For example, LFSR_S1 = K#1 **XOR** 0xffffffff; … LFSR_S15 = K#3 **XOR** IV#0. After analyzing the initialization steps of SNOW 3G cipher, we identified some weak points. For instance, the length of secret key (128 bits) is less than the length of LFSR (512 bits). The IV key includes (128 bits), the length of IV key is less than LFSR. Also, the IV key is transmitted without protection. It is more secure to use 512 bits for each of secret key and IV key, and encrypt IV key. The complexity of brute force equal to (2^{128}), if we use (512 bits for each of IV and secret key) then brute force increased to (2^{512}).

4 Analysis of ZUC Cipher Algorithm

The 4G-LTE systems used secure data transmissions via some algorithms (128-EEA3 & 128-EIA3 algorithms) [19, 20]. These algorithms designed based on ZUC cipher algorithm [21]. The ZUC needs two keys, which include 128 bits of the secret encryption key (K) and 128 bits of the Initialization Vector key (IV key). The first component of ZUC is register LFSR (S#0, S#1, … S#15). The second is Bit Reorganization (BR) (Words: X#0, X#1, X#2, and X#3). The third is nonlinear function (NLF). The NLF contains Registers R#1 & R#2, substitution boxes S-box S#0 & S-box S#1 [22].

In terms of cryptanalysis, the differential cryptanalysis method was adopted to analyze the components of ZUC algorithm. For instance, the attack focuses on the difference in values of the first byte of IV key (IV#0). To recover the secret key, the attacker explores keyspace with complexity ($2^{99.4}$) [23]. A spectral analysis method developed to attack the ZUC cipher algorithm (i.e., Version of ZUC with 256 bits). This method relied on distinguishing attacks with complexity value equal to (2^{33}) [24]. Iinitialization process of ZUC algorithm (Fig. 2) [25], includes loading register LFSR with bytes from (Secret Key: K#0, K#1, …, K#15), constant bytes (d[16]: d#0, d#1, …, d#15), and bytes from (IV key:IV#0, IV#1,…, IV#15). For example initialization process includes, LFSR_S0 = (K#0 ‖ d#0 ‖ IV#0); … LFSR_S15 = (K#15 ‖ d#15 ‖ IV#15); Where (‖: denotes concatenation).

Analysis of initialization process revealed some weak points. For instance, IV keys are transferred without protection. It is recommended to transfer encrypted bytes of IV keys. The

 https://doi.org/10.1051/itmconf/20224201004

constant values of (d[16], e.g. do= 0x44D7) used in this algorithm. It is recommended to use variable values of (d[16]). ZUC algorithm relies on secret key (128 bits), and LFSR requires 496 bits, which means secret key bits cannot cover LFSR. Complexity of brute force attack decreased from (2^{496}) to (2^{128}).

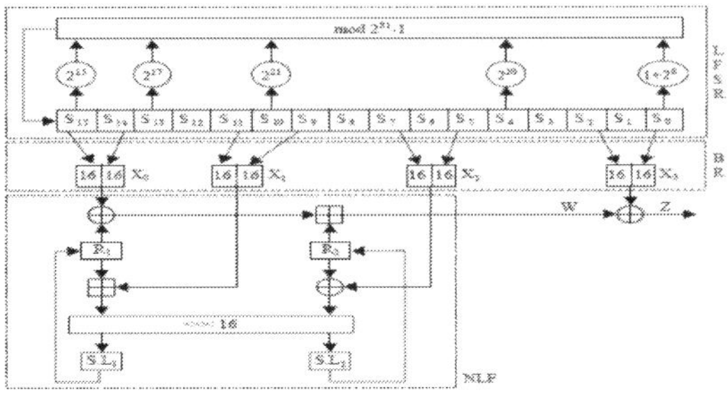

Fig. 2. ZUC Cipher Algorithm

5 Discussion

The encryption algorithms offer suitable level of data security in many applications. For instance, some encryption algorithms (e.g. RSA Enc., DES, Triple DES, and SHA256) have been used for data security in Software-Defined Network (SDN) [26]. Also, some encryption algorithms (e.g. AES, RC6, Twofish, SPECK, and chacha20) have been used for data security in Internet of Things devices (IoT devices) [27]. In this paper we focused on the encryption algorithms used in data security of 4G and 5G networks. The 4G systems (Table 1) supported Internet Accessing with Mobile Broadband networks, facilitated the connection of smart devices with Internet resources, and designed with two rigid standards (WiMAX and LTE). The 5G systems (Table 1) offered more features and relied on advanced standards. In this context, research groups focused on Dynamic Adhoc Net (DAWN), support Voice calls through IP connections (VOIP), and enhance the worldwide wireless web (wwww).

Table 1. Some Features of 4G and 5G Generations.

Mobile Technology	Features of Mobile Generations
Mobile(4G)	- Presented faster mobile broadband - Increased data capacity -Supported wider channel bandwidths (up to 20 MHz with OFDMA technique) - Offered data rates (200 Mbps, 1 Gbps in future) - Adopted multiple antenna technologies - Mobile WiMAX and LTE standards have been adopted as 4G Technology.
Mobile(5G)	- Focused on improving worldwide wireless web (wwww) - Provide voice services via IP (VOIP) - Dynamic Adhoc Net (DAWN) - Data rates (10 Gbps) - Core network (Internet), switching (All packet).

AES-CTR, this cipher depends on input and output with 128 bits. The algorithm relies on secret key K (128 bits) and IV key (128 bits). The complexity of exhaustive search of secret key is (2^{128}) (Table 2). SNOW 3G, this algorithm requires secret key (K= 128 bits) and IV key(128 bits). An exhaustive search for secret key (2^{128}) (Table 2). ZUC, the ZUC algorithm requires two types of keys (i.e., secret key K = 128 bits and non-secret IV key = 128 bits). The hacking operation of these keys requires complexity equal (2^{128}) for each of the mentioned keys. The design includes register LFSR [496 bits], BR [128 bits], and function NLF which including two registers and 2 S-boxes (Table 2). The three ciphers (i.e., AES-CTR block cipher, SNOW 3G, and ZUC stream ciphers) can be used to perform the data security in 4G-LTE and 5G generations systems.

Table 2. Symmetric Ciphers AES-CTR, SNOW 3G, and ZUC.

	Symmetric Ciphers		
	AES-CTR	SNOW 3G	ZUC
Cipher System	128-bit Block Cipher work in Stream mode	32-bit Word Oriented Stream Cipher	32-bit Word Oriented Stream Cipher
Secret Key	K=128-bit	K=128-bit	K=128-bit
IV Key	IV=128-bit	IV=128-bit	IV=128-bit
Complexity Of Initial Keys	Complexity(K)= 2^{128} Complexity(IV)= 2^{128}	Complexity(K)= 2^{128} Complexity(IV)= 2^{128}	Complexity(K)= 2^{128} Complexity(IV)= 2^{128}
Keystream	128-bit Block	32-bit Word stream	32-bit Word stream
Period of Keystream	2^{64} - 1 Where T0=64-bit	2^{512} - 1 Where LFSR=512-bit	2^{496} - 1, LFSR=496-bit
Structure of Cipher Algorithm	-Bytes Substitution (8-bit S-boxes), -Shift Rows(Byte Permutation), -Mix Columns (Linear Transform) -Add Round Key	-LFSR (S0,…, S15), -FSM(32-bit registers R1, R2, R3), and -Substitution Boxes (S1,S2)	-LFSR (16 cells), -Bit-Reorganization (128 bits), -Nonlinear Function F(32-bit memory cells R1 and R2), -S-boxes S0 and S1
No. of Rounds	10 rounds	33-steps to produce keystream	33-steps to produce keystream
Cryptanalysis Method	Improved Impossible Differential Cryptanalysis	Chosen IV Cryptanalysis	Differential Cryptanalysis
Mobile Generations	Proposed for security of: 4G-LTE and 5G	Proposed for security of: 4G-LTE and 5G	Proposed for security of: 4G-LTE and 5G

6 Conclusion

This research investigated some features presented in 4G and 5G systems. The 4G networks introduced Access to Internet resources based on Mobile Broadband, employed LTE and WiMAX standards. The design of 5G networks focused on Dynamic Adhoc Net (DAWN), Voice calls based on IP (VOIP) and worldwide wireless web (wwww). Some cipher algorithms like ZUC, AES-CTR, and SNOW 3G were presented to perform the data security in 4G and 5G networks. These ciphers possessed some characteristics. For instance, AES-CTR depends on ten rounds, 128 bits input block, and 4 math transformations. This cipher adopting secret key (128 bits). SNOW 3G algorithm employing IV and K keys. The design including LFSR register and FSM part. Brute force attack in SNOW 3G requires complexity (2^{128}). ZUC algorithm requiring IV key and secret K key. ZUC involving LFSR, BR, and function NLF. Brute force attack minimized to (2^{128}) instead of (2^{496}). For the three cipher algorithms, It is recommended to maximize the bits of initialization keys, employing encrypted forms of IV keys, and adopting dynamic variable elements of (d []) in ZUC cipher. These improvements can enhance the security of these ciphers.

References

1. Yang, J., Johansson, T., An overview of cryptographic primitives for possible use in 5G and beyond. Sci. China Inf. Sci. 63, 220301 (2020). https://doi.org/10.1007/s11432-019-2907-4
2. M. Amine, L. Maglaras, A. Argyriou, and D. Kosmanos, Security for 4G and 5G Cellular Networks: A Survey of Existing Authentication and Privacy-preserving Schemes, *J. Netw. Comput. Appl.*, vol. 101, no. November, pp. 55–82, 2018.
3. M. I. Baba, N. Nafees, I. Manzoor, K. A. Naik, and S. Ahmed, Evolution of Mobile Wireless Communication Systems from 1G to 5G: A Comparative Analysis, IJSRCSEIT,vol. 4, no. 1, pp. 1–8, 2018.
4. E. Ezhilarasan and M.Dinakaran, A Review on mobile technologies: 3G , 4G and 5G, Second International Conference on Recent Trends and Challenges in Computational Models, 2017. DOI 10.1109/ICRTCCM.2017.90
5. G. S. Nitesh and A. Kakkar, Generations of Mobile Communication, International Journal of Advanced Research in Computer Science and Software Engineering, vol. 6, no. 3, pp. 320–324, 2016.

6. S. Atapoor, Security for 4G and 5G Cellular Networks , Report for the Course Research Seminar in Cryptography (MTAT.07.022), Institute of Computer Science, University of Tartu, 2018.
7. NIST. , ADVANCED ENCRYPTION STANDARD (AES), 2001. http://csrc.nist.gov/publications/fips/fips197/fips-197.pdf.
8. M. Khan and V. Niemi, AES and SNOW 3G are Feasible Choices for a 5G Phone From Energy Perspective, Lecture Notes of the Institute for Computer Sciences, Social Informatics and Telecommunications Engineering, vol 211. Springer, Cham, 2018.
9. J. Lu, O. Dunkelman, N. Keller, and J. Kim, New Impossible Differential Attacks on AES, In: INDOCRYPT'08. LNCS, vol. 5365, pp. 279-293. Springer (2008).
10. H. Mala, M. Dakhilalian, V. Rijmen, and M. Modarres-hashemi, Improved Impossible Differential Cryptanalysis of 7-Round AES-128, In: INDOCRYPT'10. LNCS, vol. 6498, pp. 282-291. Springer (2010).
11. C. Boura, V. Lallemand, M. Naya-Plasencia, and V. Suder, Making the Impossible Possible. Journal of Cryptology, Volume 31, Issue 1, pp. 101-133, January 2018.
12. J. Park and D. Lee, FACE: Fast AES CTR mode Encryption Techniques based on the Reuse of Repetitive Data, *TCHES*, vol. 2018, no. 3, pp. 469-499, Aug. 2018.
13. G. Orhanou, S. E. L. Hajji, Y. Bentaleb, and J. Laassiri, EPS Confidentiality and Integrity mechanisms Algorithmic Approach, International Journal of Computer Science Issues, vol. 7, no. 4, 2010.
14. Specification of 3GPP Confidentiality and Integrity Algorithms UEA2 & UIA2; Doc.2: SNOW 3G specification, Tech. Specification TS 35.216 V12.0.0, 3GPP, 2014.
15. Specification of 3GPP Confidentiality and Integrity Algorithms UEA2 & UIA2; Doc.1: UEA2 and UIA2 specifications, Technical Specification TS 35.215 V12.0.0, 3GPP, 2014.
16. M. Madani, I. Benkhaddra, C. Tanougast, S. Chitroub, and L. Sieler, Digital Implementation of an Improved LTE Stream Cipher Snow-3G Based on Hyperchaotic PRNG, Hindawi Security and Communication Networks, vol. 2017, 2017.
17. A. Biryukov, Multiset collision attacks on reduced-round SNOW 3G and SNOW 3G, LNCS 6123, pp. 139–153, 2010.
18. Yang J, Thomas J, Alexander M., Vectorized linear approximations for attacks on SNOW 3G, In: Proceedings of the 27th Annual Fast Software Encryption Conference, 2020.
19. ETSI / SAGE, Specification of the 3GPP Confidentiality and Integrity Algorithms 128-EEA3 & 128-EIA3 . Document 1: 128-EEA3 and 128-EIA3 Specification, pp. 1–16, 2011.
20. ETSI/SAGE, Specification. Specification of the 3GPP Confidentiality and Integrity Algorithms 128-EEA3 & 128-EIA3. Document 2: ZUC Specification, Version: 1.6, pp. 1–18, 2011.
21. S. Lawange and M. Narnaware, Review on LTE Cryptography Algorithm ZUC and its Attacks, International Journal of Emerging Research in Management &Technology, vol. 6, Issue 7, 2017.
22. F. Nilofer and J. Qaddour, Comparative Study of Vulnerabilities in LTE Cryptographic Algorithm, International Journal of Computer Applications, vol. 180, no. 25, pp. 19–25, 2018.
23. H. Wu, T. Huang, P. H. Nguyen, H. Wang, and S. Ling, Differential Attacks Against Stream Cipher ZUC, in *Advances in Cryptology -- ASIACRYPT 2012*, pp. 262–277, 2012.
24. Yang J, Thomas J, Alexander M., Spectral analysis of ZUC-256, In: Proceedings of the 27th Annual Fast Software Encryption Conference, 2020.
25. M. Madani and C. Tanougast, Combined and Robust SNOW-ZUC Algorithm Based on Chaotic System, *2018 Int. Conf. Cyber Secur. Prot. Digit. Serv. (Cyber Secur.)*, pp. 1–7, 2018.
26. I. Aziz and I. Abdulqadder, An Overview on SDN and NFV Security Orchestration in Cloud Network Environment, *cuesj*, vol. 5, no. 1, pp. 20-27, Jun. 2021.
27. K. Jasim, R. Ismail, A. Nahi Al-Rabeeah, and S. Solaimanzadeh, Analysis the Structures of Some Symmetric Cipher Algorithms Suitable for the Security of IoT Devices, *cuesj*, vol. 5, no. 2, pp. 13-19, Sep. 2021.

ITM Web of Conferences **42**, 01005 (2022) https://doi.org/10.1051/itmconf/20224201005
ICACS21

SDN for IoT Environment: A Survey and Research Challenges

Kamaran H. Manguri[1*], *Saman M.* Omer[1]

[1]Computer Science Department, University of Raparin, Rania, Iraq,

Abstract. Recently, it has emerged that both Internet of Things (IoT), and Software Defined Network (SDN) are becoming popular technologies. The main goal of IoT is to link electronic devices via the internet, meanwhile SDN facilitates network arrangement for management of a network by distinguishing the control layer and the data layer from each other. The number of electronic devices over the internet is increasing constantly, therefore it is a complicated process to manage and control especially over the huge distributed network. IoT network can be reasonably flexible and programmable through The SDN without introducing any trouble to the previously implemented network infrastructure. This paper reviews various IoT domains and applications such as cellular network, wireless Sensor, IoT management, security and smart city framework and common IoT SDN solutions. Moreover, The IoT and SDN notion has been explored critically, with assessing the current contributions in the research field. Lastly, analyzing current available solutions for SDN-based IoT implementations comparatively helps easily understanding the emerging trends view.

Keywords: SDN, SDN Applications, IoT, Integration of SDN-IoT

1 Introduction

The Internet of Things (IoTs) is an emerging technology which enables smart ecosystem for leveraging heterogeneous technologies. Generally, in IoT network physical devices are connected to the internet such as RFID tags, actuators, wireless sensors, and/or wireless communication [1]. The IoT devices have the capability to observe, analyze and take intelligent decisions based on collected information from the surroundings and manipulation of the underlying network [2]. IoT has many applications such as unit controls, symptomatic gear, vehicles, airplane, and even atomic reactors. With regarding to secret and certain correspondences because of the nearness of the Internet of Things that provide reducing attacks and against assault. Two different models are used for security which including firewalls and system edge identification/moderation instruments to forestall outer assaults [3]. A huge data is generated and collected from IoT devices that made an issue with developers and researchers while managing, controlling, monitoring and securing of IoT devices in a heterogeneous network. In addition, heterogeneous network does not support completely in the traditional network that became an issue to fast development and

* Corresponding author: kamaran@uor.edu.krd

deployment and IoT full realization limited benefits, which are required for customers and service demands [2].

The proprietary nature of devices lead to the progress of innovation is very slow in the legacy network. Therefore, the infrastructure of traditional network and devices should be changed to realize full IoT benefits. In technologies terms, the integrated architecture may provide full IoT benefits [2]. Moreover, SDN is a new networking architecture that tries to decouple the control and data planes. This separation gives a global view of the net-work to the network controller, facilitating traffic engineering and network management at runtime [4].

In this review, different works highlighted that provide SDN based solutions for IoT environment. The recent studies during the period January 2016 - July 2021 have been reviewed by focusing on SDN applications and solutions in the IoT environment. The organization of this review is as follows. Section I provides an introduction of IoT and SDN. Background of SDN architecture addressed in Section II. Also, SDN and SDN enabled IoT architecture explained in Section III. In Section IV, the details of this review for the existing solutions are explained. The results and discussion are given in Section V. Finally, the last section remarks a conclusion.

2 SDN Architecture

The architecture of SDN includes six major components. The first component is SDN control logic which is managed by using management plane to give flexibility and easiness for implementing a new application and service. Second component is the control plane that includes one or numerous controllers and it is the most important intelligent layer. It forward various kinds of policies and rules through the Southbound Interface (SI) to the infrastructure layer. Third component, data plane represents forwarding devices on the network (routers, switches, load balancers, etc.), which is called infrastructure layer. The southbound APIs used to interact with the control plane by getting the forwarding policies and rules for applying them to the conforming devices. Fourth component, the management and control layers are allowed to communicate that are mainly a set of open source APIs using Northbound Interfaces (NI) [5]. Fifth component, east-west interfaces allows communication between various controllers that are not yet standardized. A messaging and notification system or a distributed routing protocol are used such as Border Gateway Protocol (BGP) and Open Shortest Path First (OSPF). Sixth, the SI make an interaction between data plane and control plane that may be defined such as protocols summarily, which permits push policies from controller to the forwarding plane [6].

3 SDN and SDN Enabled IoT Architecture

In SDN, control plane and forwarding plane are decoupled, and APIs are used to communicate between two planes such as OpenFlow [7]. SDN is a layered architecture that includes three layers, which are control plane/controller, application layer and data plane. Data plane layer comprises dumb forwarding devices such as switches and routers that forward data only on the controller instructions. The controller acts as a brain and controls the whole network and possesses managed by controller which provides programmability, flexible management for flow forwarding state in the data plane [1]. The application layer abstracted to the customer needs, which is Northbound used to communicate with the controller APIs such as REST full API. All applications and programs are running above the controller. The controllers since its inception are many in the market for example, Floodlight [10], OpenDaylight [7], NoX/POX, and etc. The rule of incoming flows defined via SDN controller from the data plane. SDN Layers are communicate together through open APIs called SI and NI API [1].

4 Literature Review

IoT with SDN supervision is hot topic on different platforms. Several researches and investigations are carried out to make of the using programmability SDN in IoT network.

4.1 SDN Based Cellular Network

The more complex and grouped a telecommunication environment is, the more difficult it is to obtain the satisfied outcome that needs a radio technology to be physically intervene. EOPA and RNOPA are suggested to address the reduction in SDN relayed networks with the different cloudlets in the same locations with different APs in IoT linked devices [11]. SoftAir and adds software defined gateways (SD-GWs) are used to design SDN, which are performs a cross-layer optimization between various IoT electronic devices and radio based systems during communication functionalities [12]. The scalability and availability of IoT network have been considered to address an integration IoT with 5G network because of large number of connected devices in IoT [13].

4.2 SDN for Wireless Sensor Based IoT Devices

Wireless sensors need to reduce power usage and improve the scalability in the IoT environment. For this purpose, many efforts based SDN have been seen in the literature. μSDN had proposed in [15] to solve the problems caused by applying SDN in low-power Wi-Fi networks. A novel SDN framework based WSN infrastructure are empowered to cooperate effortlessly with cloud based environments in [16]. A simulation introduced in [17] for optimization infrequent and power consumption regarding lowest control traffic.

4.3 SDN Based IoT Management

IoT management includes improving flow control and mobility in crowded and heterogeneous networks, enhancing network scalability, maintainable control, fault and exception tolerance, enforce traffic management and etc. In [18] UbiFlow suggested that is used to separate the urban-scale SDN into different geographic locations through multiple controllers. In addition, a Trust List proposed in [19] that refers to the trust distribution among IoT-linked stakeholders and appropriate combination between SDN and blockchain to enforce traffic management at the edge of networks.

4.4 SDN Security Framework for IoT

Connecting to the internet lead to open some security issues such as malware attacks and access unauthorized users. The vulnerabilities and security threats were rises that are related to IoT devices. Security Authentication is provided by the Edge servers using a lightweight authentication scheme [22]. In Addition, an utilized SDN and blockchain techniques proposed in [23] to remove the unnecessary re-authentication. Also, in [24] granular policies enforced to secure the flows in the IoT network infrastructure with providing networking authentication for IoT device.

4.5 SDN Based IoT for Smart City Application

SDN that separates data plane and control plane for the purpose of supporting configuring network compared with existing network, which is suitable for smart cities requirement, even though in the reliability faces a problem but the highest impact have showed on the system. Quality of service (QoS) aware design such as SDN-IIoT, which deals with load on the server had been proposed in [27]. QoS aware Optimum Path Selection and Rule Caching Policy (QOPS-RCP) for SDN based IoT proposed in [28].

5 Discussion

The new idea of combination of IoT and SDN is still at the beginning pace, and standardizations attempts. In spite of having multiple efforts to dominate for universal standard, the practical solution is still not available. In other words, some of the aspects are remaining conceptual/theoretical. Different kind of SDN establishment for various IoT domains with respect to the cellular network, IoT management, wireless sensor, security and smart city are given in the Table 1. It can be said that the majority of the conducted studies are not validated practically. Yet, just representative proposals framework becomes center of attraction during the recent years.

Table 1. Summary of the SDN based proposed model for IoT environment.

Application	Reference	Architecture/ Protocol/ OS	Contribution(s)
Cellular Network	[11]	EOPA and RNOPA	Produce optimum performance of reducing average cloudlet access latency and consistency
	[12]	SoftAir	Cross-layer optimization performed between IoT devices and radio based systems.
	[13]	MEC, D2D	Integrating heterogeneous IoT networks with the 5G networks and enabling dense deployment.
	[14]	NB-IoT	The proposed queueing models can be properly applied in different case scenarios.
Wireless Sensor	[15]	RPL-based IEEE 802.15.4-2012	Improving the scalability.
	[16]	WSN, MQTT	WSN infrastructure is used to develop a novel framework for the purpose of cooperating effortlessly with cloud based environments.
	[17]	-	Optimization issue is introduced using the simulation with regarding lowest control traffic.
IoT Management	[18]	UbiFlow	Suggested to use distributed hashing as supervising structure to improving network scalability and mobility controlling
	[19]	PoC	Proposed a Trust List that refers to the trust distribution among IoT-linked stakeholders
	[20]	ONOS and ODL	Developed different SDN management approach of SDN controllers.
	[21]	-	Introducing a hierarchical structure for the control plane
Security Framework	[22]	Edge computing	Proposed a framework to authorized access to the IoT devices by the Edge servers.
	[23]	Lightweight protocol	Unnecessary re-authentication is removed in repeated handover among heterogeneous cells.

	[24]	ONOS and Raspbian Virtual Machines	Granular policies enforced to secure the flows in the IoT.
	[25]	SDNWISE	A novel framework proposed for the detection of DDoS attacks.
	[26]	OVS	Proposed an efficient solutions to provide device classification and malicious traffic detection based on SDN and Machine Learning.
Smart City	[27]	SDN-IIoT and QoS	Proposed a switch application using the POX controller.
	[28]	QoS and QOPS-RCP	Proposed a QoS aware Optimum PS and RC Policy (QOP-RCP) for SDN-IoT.

6 Conclusion

The IoT seems to have a profound impact on the communication approach between human and electronic devices in the next decades. Today, the view is exceeding the limit of only connecting tangible objects via the internet. Nevertheless, as IoT is in early start of its era, it has been faced with several obstacles such as, security, programmability, flexibility, and data management. It is highly likelihood to include both programmability and centralized control in management of IoT with SDN integration. In this review the improvement of network performance with SDN has been shown for the promising solutions for current IoT. The results of these review show that solutions are not fully integrated into SDN for IoT environment. Also, most of models are evaluated and tested on the simulation models and they are not implemented on the real IoT environment. Finally, despite of many efforts had been done to build frameworks based on SDN still comprehensive architecture and framework are the challenge for researchers.

References

1. S. K. Tayyaba, M. A. Shah, O. A. Khan, and A. W. Ahmed, "Software defined network (sdn) based internet of things (iot) a road ahead," in Proceedings of the International Conference on Future Networks and Distributed Systems, 2017, pp. 1-8.
2. S. K. Tayyaba, M. A. Shah, N. S. A. Khan, Y. Asim, W. Naeem, and M. J. n. Kamran, "Software-defined networks (SDNs) and Internet of Things (IoTs): A qualitative prediction for 2020," vol. 7, no. 11, 2016.
3. A. H. Mohammed, R. M. KHALEEFAH, and I. A. Abdulateef, "A Review Software Defined Networking for Internet of Things," in 2020 International Congress on Human-Computer Interaction, Optimization and Robotic Applications (HORA), 2020, pp. 1-8: IEEE.
4. O. Salman, I. Elhajj, A. Chehab, and A. J. C. N. Kayssi, "IoT survey: An SDN and fog computing perspective," vol. 143, pp. 221-246, 2018.
5. D. Kreutz, F. M. Ramos, P. E. Verissimo, C. E. Rothenberg, S. Azodolmolky, and S. J. P. o. t. I. Uhlig, "Software-defined networking: A comprehensive survey," vol. 103, no. 1, pp. 14-76, 2014.
6. O. Blial, M. Ben Mamoun, R. J. J. o. C. N. Benaini, and Communications, "An overview on SDN architectures with multiple controllers," vol. 2016, 2016.
7. W. Braun and M. J. F. I. Menth, "Software-defined networking using OpenFlow: Protocols, applications and architectural design choices," vol. 6, no. 2, pp. 302-336, 2014.
8. J. Li, E. Altman, and C. J. Z. c. Touati, "A general SDN-based IoT framework with NVF implementation," vol. 13, no. 3, pp. 42-45, 2015.
9. J. Medved, R. Varga, A. Tkacik, and K. Gray, "Opendaylight: Towards a model-driven sdn controller architecture," in Proceeding of IEEE International Symposium on a World of Wireless, Mobile and Multimedia Networks 2014, 2014, pp. 1-6: IEEE.

10. J. Liu, Y. Li, M. Chen, W. Dong, and D. J. I. c. m. Jin, "Software-defined internet of things for smart urban sensing," vol. 53, no. 9, pp. 55-63, 2015.
11. L. Zhao, W. Sun, Y. Shi, and J. Liu, "Optimal Placement of Cloudlets for Access Delay Minimization in SDN-Based Internet of Things Networks," IEEE Internet of Things Journal, vol. 5, no. 2, pp. 1334-1344, 2018.
12. L. Tello-Oquendo, I. F. Akyildiz, S. Lin, and V. Pla, "SDN-based architecture for providing reliable Internet of Things connectivity in 5G systems," in 2018 17th Annual Mediterranean Ad Hoc Networking Workshop (Med-Hoc-Net), 2018, pp. 1-8.
13. A. A. Ateya, A. D. Algarni, M. Hamdi, A. Koucheryavy, and N. J. E. Soliman, "Enabling Heterogeneous IoT Networks over 5G Networks with Ultra-Dense Deployment—Using MEC/SDN," vol. 10, no. 8, p. 910, 2021.
14. X. Chen et al., "Traffic modeling and performance evaluation of SDN-based NB-IoT access network," vol. 32, no. 16, p. e5145, 2020.
15. M. Baddeley, R. Nejabati, G. Oikonomou, M. Sooriyabandara, and D. Simeonidou, "Evolving SDN for Low-Power IoT Networks," in 2018 4th IEEE Conference on Network Softwarization and Workshops (NetSoft), 2018, pp. 71-79.
16. C. Çeken and M. Al-Hubaishi, "Integrating SDN-enabled wireless sensor networks into the Internet," in 2019 10th IEEE International Conference on Intelligent Data Acquisition and Advanced Computing Systems: Technology and Applications (IDAACS), 2019, vol. 2, pp. 1090-1094: IEEE.
17. K. Choumas, D. Giatsios, P. Flegkas, and T. J. E. Korakis, "SDN Controller Placement and Switch Assignment for Low Power IoT," vol. 9, no. 2, p. 325, 2020.
18. D. Wu et al., "Towards distributed SDN: Mobility management and flow scheduling in software defined urban IOT," 2018.
19. K. Kataoka, S. Gangwar, and P. Podili, "Trust list: Internet-wide and distributed IoT traffic management using blockchain and SDN," in 2018 IEEE 4th World Forum on Internet of Things (WF-IoT), 2018, pp. 296-301: IEEE.
20. I. Bedhief, M. Kassar, T. Aguili, L. Foschini, and P. Bellavista, "Self-Adaptive Management of SDN Distributed Controllers for Highly Dynamic IoT Networks," in 2019 15th International Wireless Communications & Mobile Computing Conference (IWCMC), 2019, pp. 2098-2104: IEEE.
21. Z. Eghbali and M. Z. Lighvan, "A hierarchical approach for accelerating IoT data management process based on SDN principles," Journal of Network and Computer Applications, vol. 181, p. 103027, 2021/05/01/ 2021.
22. J. Li et al., "A Secured Framework for SDN-Based Edge Computing in IoT-Enabled Healthcare System," IEEE Access, vol. 8, pp. 135479-135490, 2020.
23. A. Yazdinejad, R. M. Parizi, A. Dehghantanha, and K. R. Choo, "Blockchain-enabled Authentication Handover with Efficient Privacy Protection in SDN-based 5G Networks," IEEE Transactions on Network Science and Engineering, pp. 1-1, 2019.
24. K. K. Karmakar, V. Varadharajan, S. Nepal, and U. Tupakula, "SDN Enabled Secure IoT Architecture," IEEE Internet of Things Journal, pp. 1-1, 2020.
25. J. Bhayo, R. Jafaq, A. Ahmed, S. Hameed, and S. A. Shah, "A Time-Efficient Approach Towards DDoS Attack Detection in IoT Network using SDN," IEEE Internet of Things Journal, pp. 1-1, 2021.
26. H. Gordon, C. Park, B. Tushir, Y. Liu, and B. J. a. p. a. Dezfouli, "An Efficient SDN Architecture for Smart Home Security Accelerated by FPGA," 2021.
27. H. Babbar, S. Rani, A. Singh, M. Abd-Elnaby, and B. J. J. S. Choi, "Cloud Based Smart City Services for Industrial Internet of Things in Software-Defined Networking," vol. 13, no. 16, p. 8910, 2021.
28. H. M. Mahantesh, M. Nageswara Guptha, and M. S. Hema, "Optimized Path and Reduced Rule Caching Cost for Software Defined Network (SDN) Based Internet of Things (IOT)," Wireless Personal Communications, 2021/07/03 2021.

 https://doi.org/10.1051/itmconf/20224201006

Weather Classification Model Performance: Using CNN, Keras-TensorFlow

Ashish Sharma [1*], and *Zaid* Saad Ismail [2]

[1]Department of Information Technology, College of Engineering & Computer Science, Lebanese French University, Erbil, Kurdistan, Iraq
[2]Department of Accounting, College of Administrative and Financial Science, Knowledge University, Erbil, Kurdistan, Iraq

Abstract. Nowadays, automation is at its peak. The article provides a base to examine the weather through the machine without human intervention. This study offers a thorough classification model to forecast a weather type. Here, the model facilitates defining the best results for the weather prediction model to any climatic zones and categorizes the climate into four types: cloud, rain, shine, and sunrise. This model designs and reveals using convolution neural networks (CNN) algorithms with Keras framework and TensorFlow library. For practical implementations, use the images dataset available from the kaggle.com website. As a result, this research presents the performance of the designed and developed model. It shows accuracy, validation accuracy, losses, and validation losses approximately 94%, 92%, 18%, and 22%, respectively.

Index Terms: Climate, Convolution Neural Network (CNN), Keras, TensorFlow, Weather Classification Model.

1 Introduction

This article represents the weather classifications model without human intervention. The paper introduces an intensified climate foresight model that employs the notion of data mining in weather prediction. The model can divine any characteristic feasible in the dataset while other data mining-based algorithms produce particular traits. It is computationally valuable and executes appropriately to mobile devices like the Android ecosystem [1]. It uncovers the distinct climatic conditions and separates the cloudy, rainy, shiny, and foggy patterns. Here, the genetic classifications comprise based on the geographic determinants of the environment and the surface energy resources, plus concerns of the air-mass review [2, 6].

It introduces a system that supports a regional weather type class procedure to aid interpretation in city atmosphere considerations. Here, this helps analyze the change in particular environmental conditions from hot to cold or cold to hot, high to low, or low to

* Corresponding author: ashish.sharma@lfu.edu.krd

increased wind speed. As it happens frequently, it intends to define future observations appropriately [3]. It proves climate change peculiarity technically on the baseline high-resolution climatic projections. Both periods measure confidence levels of the assemblage range, presenting relevant implications regarding the authenticity, like the accuracy of the classifications [4].

This system fulfills probability density functions (PDF) to perceive and simulate temperature and precipitation according to zones for various climates. The research examines temperature and precipitation twists above current weather results and identifies systematic errors of the model to acquire better performance. This system comprises many models from different regions to achieve accuracy with results [5].

Machine learning techniques are worthwhile in developing real-time applications, including computer vision, image processing, speech recognition, text processing, Etc. Moreover, it provides beautiful results for classifying text data according to distinct climatic conditions like temperature and humidity [7].

A convolution neural network (CNN) is an artificial neural network (ANN) specified for image pattern detection. It is suitable for image analysis to identify the features of an image, called feature maps. CNN is composed of hidden layers are known as convolutional (ConV) layers. These ConV layers accept input, transform input into some mode, and convert it to output. Now the same production is information to the next layer. This modification is a convolution operation that facilitates detecting patterns from images precisely and accurately. Each convolution layer needs to specify the number of filters. The filters play an essential role in detecting edges called "edge filters" and detecting objects called "object filters." So, the CNN technique is adopted for this proposed work to identify the pattern of the weather images dataset [13].

This work investigates the performance of the designed and developed deep learning model by using CNN architecture with Keras framework and TensorFlow system-defined libraries for Classification on the original Kaggle.com website weather image dataset. Finally, these deep learning algorithms work for two primary responsibilities: Classification and predicting climate situations.

2 Literature review

The author studied Kappen's climate classification schemes. It facilitates regarding analysis of weather changes according to regions and climatic cycle time. Here due to doubts among the classification models to identify variation in the climate zone with measurements and forecasts until outcomes are unclear [12].

The scholars described pressure Laplacian and the geopotential Laplacian algorithms for climatic classifications and represented the output with the help of the regional climate models (RCM) simulation model. Any severe environmental conditions like dryness, floods, high-speed wind, heavy snowfall, Etc., harm biodiversity [11].

The researchers described that global weather types understanding is a great challenge. Manually categorizing different weather types as per temperature and precipitation is difficult to confirm for different longitude, latitude, altitude, Etc. It is an arduous task due to the availability of data resources according to areas. Consequently, mentioned above, severe climatic conditions support further research in weather classifications [8, 10, 11].

The authors compared several high performable machine learning models on multi-class datasets. The various systems developed with different algorithms such as decision tree CART, gradient boosting, KNN, linear regression, lasso, ridge, MLP, deep learning, SVM, and random forest achieve accuracy for weather classification. These models are showing some overfitting. As per the testing phase, the linear regression technique is highly performable for a prediction model. Furthermore, the k- nearest neighbors algorithm is lower in performance [7].

The further authors explained the spatial augmentation of a gridded weather typing classification (GWTC) design to a global rule. Several methods apply for making the system robust in terms of weather classification to recognize the severe climatic conditions [6].

The scholars outlined new weather datasets correlated with the present experimental setup to simulations of REMO from the CORDEX structure. These simulators tend to analyze the climate classifications. The REMO produces relatively large yearly precipitation and temperature inclinations in dry and arctic zones and low skill. The biased outcome always represents an uncertainty to measurements of any weather data result [5].

This research focuses on biodiversity problem-solving strategies. The author studied for one (01) KM region on the map and investigated weather changes. The results presented with higher accuracy and climatic conditions are suitable for vegetation, spices [2, 4].

Moreover, the intellectuals examine the weather classification through the local weather type (LWT) approach. This method is getting direct based on atmospheric data and statistical techniques. It facilitates an analysis of the atmospheric parameters near the ground. It is a proven procedure to scrutinize future problems before time [3].

The researchers observed the difference between k-nearest neighbors' (KNN) and modified KNN (MKNN) for weather classification. Here it shows that the MKNN provides more reliable results than other datamining projection ways [1].

3 Methodology

3.1 Algorithm for Weather Classification Model

This algorithm accomplishes ten steps process to design and develop a weather classification model. These are the following steps:

[1]. Initiate a model
[2]. Import all of the required system defined libraries
[3]. Preparation of image dataset for training the machine and validating the machine model
[4]. Show graphically for classifying the training and validating data into distinct classes
[5]. Build a model using CNN
[6]. Compile a model
[7]. Train a model as per the training dataset
[8]. Validate a model as per the validation dataset
[9]. Display the performance of an implemented model graphically
[10]. Finish the process

3.2 Pseudo-code of the proposed algorithm

The pseudo-code of the proposed algorithm for a weather classification model, such as:

Step-1. *Initiate a model.*

Step-2. *To import all the required inbuilt libraries, such as :*
TensorFlow, Keras, NumPy, Matplotlib

Step-3. *To prepare data for training the model and validating the model.*

- *Set path for image data collection*
- *Classify the image data into four classes: Cloud, Rain, Shine, and Sunrise.*

- *Divide image data into distinct batches, i.e., TRAINING_DIR and VALIDATION_DIR.*

Step-4. *Display bar-graphs*

- *Plot bar-graph for Classification of the training dataset.*
- *Plot bar-graph for Classification of the validating dataset.*

Step-5. *Build a new Weather Classification model by CNN.*

- *Design a sequential model with convolution 2D filters, kernel-size, activation function 'relu,' and padding function 'same'; maximum pooling size of 2D images; add a dense layer with activation function' softmax.'*
- *Summarize the model.*

Step-6. *Compile a new Weather Classification model.*

Step-7. *Train a new Weather Classification model as per the collected images dataset.*

Step-8. *Validate a new Weather Classification model as per the collected images dataset.*

Step-8. *Plot the line graphs to show the performance of 'Training and Validation Accuracy' and 'Training and Validation Loss.'*

Step-9. *End.*

4 Results and discussion

The proposed implemented pseudo-code outcome for the weather classification model is step-by-step performance. The resultant of the proposed model contains as:

4.1 Prepare data

The initial phase prepares the image dataset for further processing with the implementing model. Here, the data details are as follows. See Figure-1 to clarify the details of each type of image for Train and Test (Validation) Datasets.

```
Train set Cloud 240
Train set Rain 173
Train set Shine 203
Train set Sunrise 287

Test set Cloud 60
Test set Rain 42
Test set Shine 50
Test set Sunrise 70
```

Fig. 1. Prepare dataset.

4.2 Exploratory data analysis

It categorizes the data into four classes: Cloud, Rain, Shine, and Sunrise. See Figure-2.

```
['Rain', 'Shine',
'Sunrise', 'Cloud']
```

Fig. 2. Exploratory data analysis.

4.3 Classification of training and validation datasets

After preparing the dataset, it classifies into distinct parts as per the model's requirement. This practice helps in training and then validating the proposed weather classification model. Here, it represents the data in a bar graph as per class. Each bar shows different data sets like cloud image data, rain image data, shining image data, and sunrise image data for training and validation datasets. It shows in Figure-1 for Classification of Training Dataset and Classification of Validation Dataset, respectively.

4.4 Implementation of weather classification deep learning CNN model

This proposed sequential weather classification model architecture comprises eighteen layers: nine Convolution, five MaxPooling, one Flatten, and three Dense layers. This designed model extracts all of the features from dataset images. Accordingly, train the model as per extracted features. It is more and more effective with cleaned data. Thus, it defines that provide the best clean images data to model and get high performing results. Hereafter training, it expresses, as a result, all of the parameters that are 10,411,108 trained. And no single parameter without training. It shows the efficiency of the model.

4.5 Compile and train the model

After developing the model, compile the model to check all types of syntax errors. Now train the model with the collected clean dataset by following thirty (30) epochs or repetitions to get better results.

4.6 Analysis of the model's performance

This step displays the outcome in 'Training and Validation Accuracy' and 'Training and Validation Loss.' It shows accuracy, validation accuracy, losses, and validation losses approximately 94%, 92%, 18%, and 22%, respectively, as look in Figure-3 and Figure-4. It is pretty explained here is no overfitting in the line graphs. All four curves show similarities for training and validation of an implemented model. It is a great performing model.

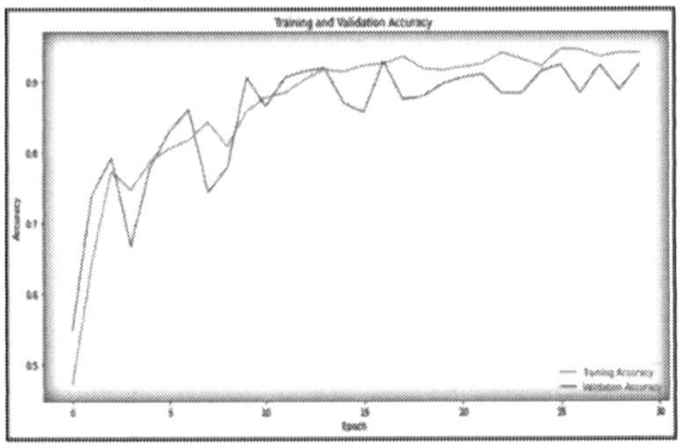

Fig. 3. Training and validation accuracy for the weather classification model.

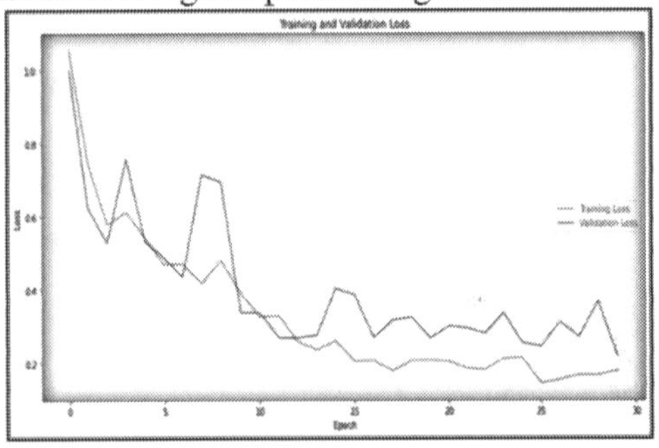

Fig. 4. Training and validation loss for the weather classification model.

4.7 Confusion matrix of the model's performance

This confusion matrix shows the results for total dataset images of 903 are [[57, 48, 73, 62], [37, 36, 42, 58], [54, 45, 47, 57], [51, 72, 66, 98]]), such as correct prediction for cloud, rain, shine, and sunrise are 57 out of 240, 36 out of 173, 47 out of 203 and finally 98 out of 287, respectively, look in Figure-5.

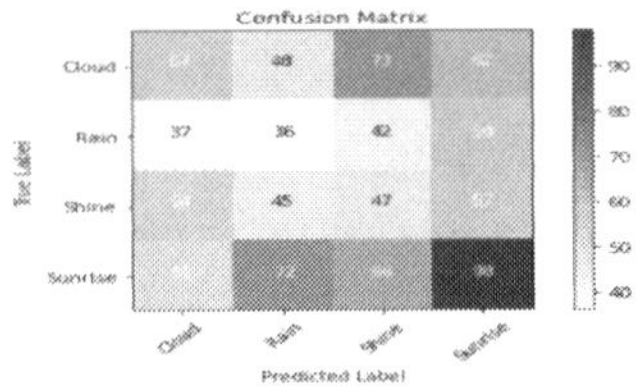

Fig. 5. Confusion matrix

5 Conclusion and future work

This work concludes the implemented weather classification model with its performance. After following all of the processing steps, it shows incredible results. It here implemented model trained through collected dataset from Kaggle.com website. This image dataset is a collection of clean data. It finds 94% training accuracy, 92% validation accuracy, 18% training losses, and 22% validation losses. As well as no overfitting or underfitting notice during the training and validating stages. So, it concludes that the model performance is highly acceptable.

In the future, this model will extend with more attributes like humidity, precipitation, air pressure, atmospheric pressure, temperature, solar radiations, Etc. and then integrate with an early warning system. It will provide the facility of finding natural disasters before time plus support to predict global warming in real-time.

References

1. M. Abrar, A. Sim, D. Shah, S. Khusro, A. Salam, *Weather Prediction using Classification*, Science International. **26**, 2217-2223, https://www.researchgate.net/publication/272482887_Weather_Prediction_using_Classification, (2014)
2. A. John Arnfield, *Climate classification*, Encyclopedia Britannica, https://www.britannica.com/topic/classification-1703397, (17 Mar. 2016)
3. J. Hidalgo, R. Jougla, *On the use of local weather types classification to improve climate understanding: An application on the urban climate of Toulouse*, PLoS ONE 13**(12)**: e0208138. https://doi.org/10.1371/journal.pone.0208138, (2018)
4. H. Beck, N. Zimmermann, T. McVicar, et al., *Present and future Köppen-Geiger climate classification maps at 1-km resolution*, Sci Data **5**, 180214. https://doi.org/10.1038/sdata.2018.214, (2018)
5. Armelle R. Remedio, C. Teichmann, et al. *Evaluation of New CORDEX Simulations Using an Updated Köppen–Trewartha Climate Classification,* Atmosphere 10, no. **11**: 726, https://doi.org/10.3390/atmos10110726, (2019)
6. CC. Lee, *The gridded weather typing classification version 2: A global-scale expansion*, Int J Climatol. 2020, **40**: 1178– 1196. https://doi.org/10.1002/joc.6263, (2020)
7. I. Gad, H. Doreswamy, *A comparative study of prediction and classification models on NCDC weather data*, International Journal of Computers and Applications. 1-12, 10.1080/1206212X.2020.1766769, (2020)
8. A. Gardner, I. Maclean, K. Gaston, *A new system to classify global climate zones based on plant physiology and using high temporal resolution climate data*, Journal of Biogeography. 47. 10.1111/jbi.13927, (2020)
9. K. Piotrowicz, D. Ciaranek, *A selection of weather type classification systems and examples of their application*. Theoretical and Applied Climatology. 140. 10.1007/s00704-020-03118-2, (2020)
10. Y. Lai, N. Tanaka, S. Im, K. Kuraji, et al. *Climate classification of Asian university forests under current and future climate*, Journal of Forest Research. **25**. 1-11. 10.1080/13416979.2020.1759898, (2020)
11. A. Belušić Vozila, M. Telišman Prtenjak, I. Güttler, *A Weather-Type Classification and Its Application to Near-Surface Wind Climate Change Projections over the Adriatic Region*, Atmosphere, 2021; 12**(8)**:948. https://doi.org/10.3390/atmos12080948, (2021)
12. D. Cui, S. Liang, D. Wang, *Observed and projected changes in global climate zones based on Köppen climate classification*, WIREs Climate Change, 12. 10.1002/wcc.701, (2021)
13. Ketkar N. Convolutional Neural Networks. In: Deep Learning with Python. Apress, Berkeley, CA. https://doi.org/10.1007/978-1-4842-2766-4_5, (2017)

A Framework for Optimizing Software Regression Test Case based on Modified-Ant Colony Optimization (M-ACO)

Abrahmi Silvarajoo[1]*, *Mohammed* Hazim Alkawaz[2]

[1]School of Graduate Studies Sciences, Management and Science University, Shah Alam,Selangor,.Malaysia
[2]Faculty of Information Sciences and Engineering, Management and Science University, Shah Alam,Selangor,Malaysia

Abstract. Software Regression Testing (SRT) helps with the modified code to ensure that no other new defects have been trained and significantly impact the existing code. SRT is a time and cost taking process. Test Case Prioritization (TCP) helps to reschedule test cases (TC'S) and picks according to the prioritization, which allows the software testers to maximize the number of faults detected within a given time frame. Ant Colony Optimization (ACO) is an optimization algorithm motivated by the searching behavior of insects. This paper proposes a simple framework for the novel Modified-ACO (m-ACO) to prioritize the TC's. m-ACO is planned as the adjusted type of the ACO method, which reschedules the execution sequence of TC'S by changing the wonder of characteristic insects for choosing their food source. The proposed Framework will help the software testers save their time to execute SRT TC's. It has been tested for Zasta Billing Web Application and compared the time taken to complete the SRT manually and using the proposed Framework.The proposed Framework helps to reduce the time taken from 90 days to 45 days to achieve the SRT.

1 Introduction

ST measures the quality of the software [1]. The focal point of quality is to find and remove defects [2]. Inside the test cycle, the testing procedure that guarantees that the highlights are not influenced by the progressions made in programs is known as SRT [3]. The objective of SRT is to ensure that any new adjustments of the product ought not unfavorably influence the current capacities, and there is no further shortcoming covered up in the embedded code [4].SRT is regularly performed throughout the programming improvement and upkeep life cycle [5].Nonetheless, as the product turns out to be huge, the quantity of the test cases will increment after some time, and the expense of performing SRT could turn out to be increasingly costly[6]. An enormous scope ST, a complete SRT frequently takes weeks or even a long run [7]. For this situation, the tester needs to sort the TC's to ensure that the higher priority can be executed right on time as expected under the circumstances [8]. This paper

* Corresponding author: abrahmisilvarajoo@gmail.com

proposes a framework for TCP's novel technique, "m-ACO." Research has been conducted for previous m-ACO studies, and we have come up with a user-friendly framework for m-ACO, which will help the software testers prioritize the TC's very quickly. Also, previous studies only focused on generating the algorithm and executing it in a small SRT environment[9].But, in this study, the proposed algorithm has been discussed in detail. It has been tested for a startup company web application, Zasta Billing Web Application. m-ACO is planned as the adjusted ant colony optimization method, which reschedules the execution sequence of SRT TC'S by changing the wonder of characteristic insects for choosing their food source.

2 Literature Review

SRT is one sort of ST performed with current software changes [10]. It is crucial to give conviction that such changes, which areas of late introduced, don't impede the lead of the unaltered existing bits of the software. Retesting all,TC Selection (TCS),TCP,and TC Reduction (TCR)/TC Minimization (TCM) are regression testing activities[11-12].Retest all is a re-execution of all TC's in a specific capacity test suite. It is an excellent skill, yet be that as it may, it can devour a great deal of time. Instead of re-executing the whole test suite, regression TCS is intelligent to choose a specific piece of a test suite to be run. TCP is the strategy to organize the rundown of TC's, empowering TC's with higher need, chosen based on some predefined paradigm, to be executed before the experiments with a lower need, to meet the required execution objective [13]. So TCP strategy does not dispose of any TC's, thus keeping away from the downside of the TCM procedure.TCM procedures eliminate the repetitive TC's to limit the number of TC's in a test suite. The hybrid approach is a combination of TCS and TCP. Different prioritization criteria might be connected to the regression test suite to meet a specific measure. TC's can be prioritized as far as random, optimal, statement coverage total or additional, branch coverage total or additional, failure rates, or total fault, exposing the TC's potential [14].

The ACO method fundamentally depends on the insightful strategy utilized by the ants to discover the food source [15]. ACO is an optimization algorithm for distinguishing proof of optimal balance between different factors to limit the search space in an inevitable derivation of the final solution. It is a metaheuristic swarm intelligent-based tool that imitates the conduct of ants searching for sustenance. Ants can discover the optimized way to specific nourishment source from their home by indirect interaction with every other individual from their colony, for example, sending a message to all other members of their colony through environment [16]. This marvel is known as stigmergy. The decision procedure of ants is impacted by the pheromone trails saved by other ants on the ground since ants proceed onward the way, which has a more elevated amount of pheromone when contrasted with different methods.

This paper proposes a framework for an m-ACO, which eases the software testers to perform their SRT.m-ACO,is a changed ACO structure.When real ants discover a nourishment source, it returns with sustenance to their nest, store the food in the nest and afterward go to a similar nourishment source again until the nourishment source wraps up.
On the off chance that that way is most brief, at that point, inevitably, all ants will begin following a similar trail until that sustenance source wraps up and it declines the diversity of the kept nourishment in a given time.The previous studies only focused on developing an algorithm without a proper framework explanation, and the developed algorithm does not get implemented well. Thus, this paper will help researchers to have an appropriate view of the Framework for the m-ACO algorithm, which helps prioritize the TC's using the Average Percentage Faults Detection (APFD) metric. It has been implemented for a project in a startup company Zasta Infotek.

3 Methodology

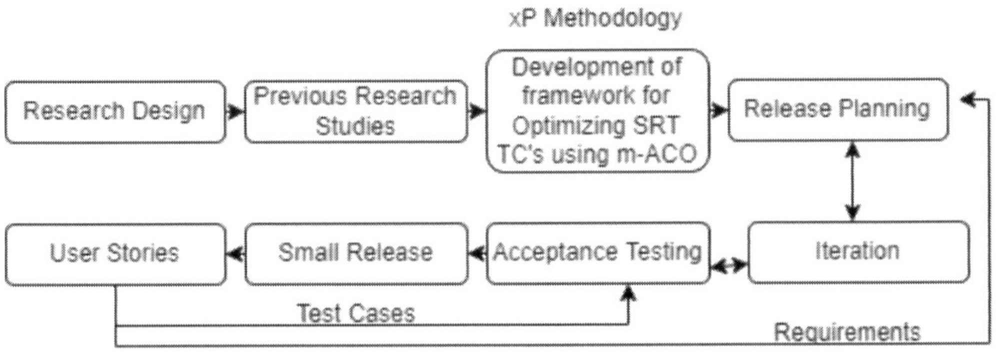

Fig. 1. Research Methodology

Figure 1 depicts the complete diagram of the research methodology, which starts from designing the research to completing the development of the proposed m-ACO framework. The current research was conducted in a few phases, as detailed in the following:

1. The first phase includes studying current RT problems, previous research studies-related to the proposed framework, gathering information on SRT, TCP, and forming the Literature Review.
2. The second phase includes collecting data to develop the proposed framework by conducting an unstructured interview with ST Project Manager from startup company Zasta Infotek Private Limited based from India.
3. The third phase includes developing the proposed framework for optimizing SRT TC's using the M-ACOa mechanism using XP methodology.
4. The fourth phase includes testing the proposed framework through acceptance testing by performing experiment analysis for test suites developed for a web page called Zasta Billing Web Application.
5. The fifth phase includes the discussion of the results obtained from the experiment analysis and the formulation of both the conclusion and recommendation

The type of research methodology that has been used is the quantitative research method which is experimental research.It has been conducted for Zasta Infotek Private Limited, India, and tested for a project called Zasta Billing Web Application(ZBWA). ZBWA is an online invoicing application for small businesses, freelancers, and accountants. The framework has been developed using Angular-typescript. Next is the use of Firebase. It helps to store the database for the framework. A laptop/PC with at least 4GB RAM has been used to develop this framework.

4 Results

The framework has been successfully developed. The result has been collected by comparing the time taken for SRT conducted for Project ZBWA in Zasta Infotek Private Limited, based in India.SRT has been carried out for ZBWA using the developed framework and compared the time taken for SRT with the M-ACO framework and without it. The time to complete the SRT using the M-ACO framework is 45 days, whereas without M-ACO is 90 days. The screenshots explain the main framework functionalities.

The testers can upload, view, and directly calculate the time taken by uploading the test case report by filling up the necessary columns. The tester has two options to choose for the upload type. When the tester chooses the new upload type, they can fill up the report name, select the iteration number, and upload their report. Once the details have been filled in, they can submit them to prioritize their TC's. When the tester chooses the existing upload type, the tester selects the report name, the iteration number and uploads their document. Once the details have been filled in, they can submit them to prioritize their TC's. The tester needs to fill up some mandatory information in the template such as TC number, description, priority, type, status, created date, and execution time. The tester can upload either new or existing TC documents with the stated mandatory fills as above.

Fig.2. Example of updating existing document

When the tester chooses an existing TC's document, the iteration number, Total TC's, TC passes, and failure will be automatically filled according to the TC document that has been uploaded. After all the necessary details are filled in, the tester can submit the details. The tester can search and view theTC report that has been submitted in the view component. To calculate the priority, the tester needs to choose theTC report submitted previously.

Fig.3. Test Case Priority Calculation

The priority determined based on time-dependent rate computation was utilized as an approval boundary. The m-ACO strategy for TCP determined the appropriateness of every node based on optimal values of the code covered, faults detected, and time utilized for execution. A new report with a prioritized list will be generated. The artificial ants in a pseudo-random way moved toward the fault-carrying modules and determined the appropriateness. A pheromone factor was scored to draw in other antlike measures. Just those pheromone trails that had a place with the most practical code module became more grounded in later cycles by progressive appearances of more antlike actions. The pheromone trail of less-visited modules became more vulnerable by a consistent rate. Ants might, in any case, visit these progressively weakening paths, however, with lower likelihood. Two parameters were determined: Average percentage of faults detected and Percentage of Test Suite Required for Complete Fault Coverage (PTR). The primary measure utilized for focusing on test cases in this proposed m-ACO method depended on accomplishing total or most excellent fault coverage. Then the outcomes were investigated utilizing the APFD metric [17]. The APFD metric is used to assess the average fault detection rate per test suite execution level.

$$APFD = \frac{1 - TF1 + TF2 + TF3 \ldots\ldots TFm}{mn} + \frac{1}{2} \quad (1)$$

The notation for APFD calculations are,T: test suite to be evaluated,m :number of faults in an application under test,n:total test cases in a test suite,TFj: location of the first test case in test suite T that reveals fault j.

5 Conclusion

SRT is the confirmation of beforehand functional software regarding whether it remains so after a change. It is a time and money concentrated interaction.TCP is beneficial and a sensible RT procedure to limit the number of TC's in a test suite by rearranging TC's in need request, which viable builds the likelihood of code coverage and pace of fault discovery through the APFD metric. This paper presented a framework for a new algorithm, m-ACO, an altered form of ACO for the TCP. m-ACO works by changing the procedure utilized by regular subterranean insects to determine their food source. Research has been conducted for previous m-ACO studies, and this paper has come up with a user-friendly framework for m-ACO, which will help the software testers prioritize the TC's very quickly. Also, previous studies only focused on generating the algorithm and executing it in a small SRT environment. But, in this study, the proposed algorithm has been implemented for a SRT Project for a startup company.The future work on the paper will focus on developing the framework as a mobile application to make the testers ease their testing activities.

References

1. Iboyi, V, *Software Quality: A Case for Regression Testing,* (2020)
2. Kaur, S, *A Review of Software Development Life Cycle Models.*, International journal of advanced research in computer science and software engineering **5(11)**: 354-360,(2015).
3. Dahiya, O. and K. Solanki,*A Systematic Literature Study of Regression Test Case Prioritization Approaches,* International Journal of Engineering & Technology **7(4)**: 2184-2191, (2018).
4. Bajwa, J. K. and R. Kaur, *An Adaptive Approach for Test Case Prioritization in Regression Testing using Improved Genetic Algorithm*, (2017).
5. Dalal, S. and K. Solanki,*Challenges of Regression Testing: A pragmatic perspective,*International Journal of Advanced Research in Computer Science **9(1),**(2018)

6. Greca, R.,*Orchestration Strategies for Regression Testing of Evolving Software Systems* ,(2020).
7. Gao, D., et al.,*Test case Prioritization for Regression Testing based on Ant Colony Optimization,*2015 6th IEEE international conference on software engineering and service science (ICSESS), IEEE,(2015).
8. Khari, M., et al.,*Test Optimisation: An Approach based on Modified Algorithm for Software Network,*International Journal of Advanced Intelligence Paradigms **17(3-4)**: 208-237,(2020).
9. Hasnain, M., et al.,*Functional Requirement-Based Test Case Prioritization in Regression Testing: A Systematic Literature Review,*SN Computer Science **2(6)**: 1-32, (2021).
10. Ali, M. and U. Khan, *A Survey Paper, Test Cases Prioritization of Regression Testing,* (2021).
11. Kiran, R.,S..,*A literature survey on TCP-test case prioritization using the RT-regression techniques,* Global Journal of research in engineering, (2015)
12. Kumar, A. and K. Singh,*A Literature Survey on Test Case Prioritization.*" Compusoft **3(5)**: 793.,(2014).
13. Kaur, A., *An Approach to Extract Optimal Test Cases Using AI.* 2020 10th International Conference on Cloud Computing, Data Science & Engineering, IEEE. (2015)
14. Kumar, S. and P. Ranjan,*ACO based test case prioritization for fault detection in maintenance phase,*International Journal of Applied Engineering Research **12(16)**: 5578-5586.,(2017).
15. Alkawaz, M.,H. and A. Silvarajoo, *A survey on test case prioritization and optimization techniques in software regression testing,*2019 IEEE 7th Conference on Systems, Process and Control (ICSPC), IEEE,(2019).
16. Solanki, K., et al.,*Impact of m-ACO-Based Test Suite Prioritization on Software Quality,* Applications of Advanced Computing in Systems, Springer: 289-297.,(2021)
17. Solanki, K., et al.,. *Test case prioritization: an approach based on modified ant colony optimization (m-ACO).*2015 International Conference on Computer, Communication and Control (IC4), IEEE.,(2015).
18. Solanki, K., et al.Test *case prioritization: An approach based on modified ant colony optimization.* Emerging Research in Computing, Information, Communication and Applications, Springer: 213-223.,(2016).

Artificial Intelligence Effectiveness and Impact within COVID-19

Ragad M Tawafak[1*], *Ghaliya* Alfarsi[1], and *Baidaa Hamza* Khudayer[1]

[1]Department of Information Technology, AlBuraimi University College, Buraimi, Oman

Abstract. Artificial intelligence is one of the most important programs that are used in all sectors of society. Accordingly, artificial intelligence has become required to help combat Covid 19. In the field of education, artificial intelligence has become the ideal alternative to study and complementary to distance study. This research provides a summary of the types of artificial intelligence, its benefits, drawbacks, and the field of its uses. This paper will present the importance of using intelligence Artificial intelligence in the field of education specifically and its importance in expanding students' awareness in the case of distance study.

1 Introduction

Artificial intelligence is being used professionally to combat the COVID-19 pandemic [1] after the occurrence of infections and the complete closure all over the world. During the difficult period [2] of the pandemic, the most common solution has been the use of artificial intelligence. Artificial intelligence has invaded all different aspects of life. This study focuses largely on the applications of artificial intelligence in the educational aspect to simulate the new method of education within the period of Covid 19. [3, 4].

Artificial intelligence has been used in the field of learning with multi-dimensional moving images. Where imaging takes place, many different moving images appear to reflect the opportunity of smart devices or robots [5]. These robots help describe complex objects by drawing detailed maps of the mechanics of objects and how they are made. They made machines to help them deliver more within the processing plants. Fortunately, this idea has generally been successful, and individuals still use dedicated devices to achieve great speed and high cost [6, 7].

This research is divided into several units. It begins with an introduction and general review of research using artificial intelligence in education and in motion pictures to serve learners. Next, an explanation of the types of artificial intelligence in the areas of life. Then, the benefits of artificial intelligence and some of the qualities that are obvious as drawbacks in its uses. In the end, the research refers to the summary of the paper and the supported recommendations for the use of artificial intelligence in the field of education and distance learning.

* Corresponding author: raghad@buc.edu.om

2 What is the necessity of Artificial Intelligence and its features?

In this recent period, the measure of knowledge produced by two people goes beyond the plurality of robots and smart applications that help in the speed of completion and accuracy of information. [8, 9, 10]. In gaming, far fewer people would see excellent checkers bosses, with over 500 x 1018, or 500 quintillions, of various possible moves [11]. Personal computers are incredibly adept at checking these algorithms to indicate a more direct identification. [12]. Privacy and Security: Licensors can determine the effectiveness and efficiency of security and privacy settings. Protection is based on the concept of "use". It is to give powers to the agency that prepares the program and to enable it to freely add to external parties working on development and assistance [13]. Mutual Suspicious "memory-free" subsystems are implemented easily and efficiently. The Protection Officer has the right to direct access to menus relating to controls, objects, and capabilities. Matters related to protection always remain unresolved until the issue of trust and intent becomes clear from the party dealing with it [14].

There are a lot of data that stimulate the desire for appropriate evaluation methods for machines capable of performing complex tasks [4, 5], including the measure of scientific progress in the field of study: the degree of research progress, the percentage of variation of copies in the same system, identifying obstacles to modification and switching, between the components of a system One or more AI systems with different tasks [10]. The goal is to determine the variables that take into account speed, accuracy, reliability, independence, the reasons deduced from the relationships, the degree of complexity of things, and other factors that affect the construction of the smart system [15].

In the modern period of technological development, educational methods focused on the ability to play smart games by students as an indicator of the development of systems. Dealing with a variety of smart tools and algorithms to achieve the conditions of the game. Hence simple performance A rating such as the achieved game score cannot describe the progress of AI per se [6]. Since the general assessment of the player's ability to control the data and change the various characteristics, we can conclude the degree of the player's self-control.

3 Features– The New Age Electricity

Eliminate boring and cumbersome processes [15].

Absorbing large and compressed information [16].

It imitates human perception in the speed of discrimination and reactions, for example, thwarting catastrophic events, the accuracy of facial recognition, chatbots, environmental protection speed mechanisms, the development of technology for the use of daily life, and others [17].

4 Advantages and Disadvantages of Artificial Intelligence

The positive aspects of synthetic knowledge applications are enormous and could work for any specialist department [18, 19]. What do you think that we see a number of them?

4.1 Reducing human error

The emergence of this term is because people make mistakes from time to time [1, 9]. These errors are simple and comprehensible, and some are complex and linked to other systems that negatively affect several delicate systems. Conversely, computers do not make these errors

if they are properly customized [2, 12, 20]. As the synthetic knowledge mechanism, it is done by using the alternatives taken from the recently accumulated data from real and reliable sources by applying the chosen order of the calculations. So, errors are automatically reduced, so the possibility of reaching accuracy with a detectable level of accuracy may be an opportunity [2, 12].
For example: When anticipating the use of AI, they need to reduce most human errors and human use as possible and replace them with sophisticated robots.

4.2 Takes risks instead of humans

Building and developing robots that are used specifically for rescue operations or to use in areas that are not safe for humans. For example: defusing the bomb, inspecting the deepest parts of the seas, drilling for coal and oil, analyzing and prior detection of infected people, especially in the Covid-19 period, and tending to be used successfully in any form of natural or man-made disasters [2, 3].
Example: Have you ever been exposed to the winds of an explosion at the Chernobyl atomic power plant in Ukraine? At that time, there were no artificial intelligence robots that could help us reduce the impact of radiation by controlling the stove ahead of time since any human who gets close to the middle dies very quickly [4]. In the end, they poured sand and boron from helicopters from a small interval [5]. Figure 1, shows a basic algorithm and an advanced one to find a smart solution.

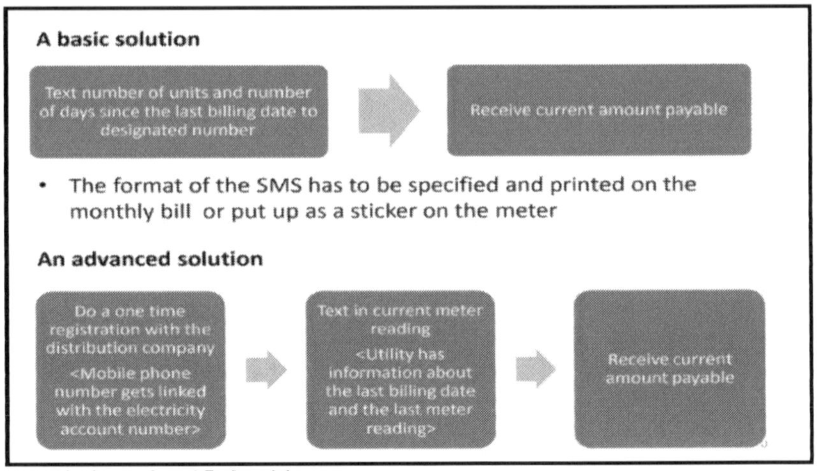

Fig. 1. Smart solution using AI algorithms

4.3 Available 24 x 7

The daily and normal work of the average person ranges between 4-6 hours a day, excluding rest periods. During the break, people work to motivate themselves and prove themselves against another working day. But in the case of using robots and artificial intelligence software, unlike people, we will make the machines work 24/7 without breaks and will not be stopped, making the work output double the original work if it was under human management [2, 6].
Model: Educational institutes and help centres are now providing solutions to many questions and issues that can be addressed for the successful use of AI through algorithms that discover frequently asked questions and hypothetical solutions for each case. Figure 2, shows the 24x7 camera working with AI development.

Fig. 2. AI Camera for data analysis algorithms

4.4 Digital assistance

Organizations and associations rely on ready-made AI-related software to interact with customers to reduce the need for human resources. Some chatbots were planned so that it would be difficult to verify our visit using a chatbot or personal software [7, 8, 9].
Form: We all understand that associations have a customer assistance group that requires clarifying customer questions and inquiries. Using artificial intelligence, associations can create a voice bot or chatbot that may help customers respond to each of their problems. We will see many associations that have already started using it on their various websites and apps [3].

5 Disadvantages

5.1 High construction costs

Due to the continuous renewal and updating in this field related to artificial intelligence, and the mechanism used to clarify the curricula in the case of distance learning, as is required especially in the medical field and the accompanying development of diagnostic software for Covid 19. Equipment and programming must be updated over time to meet the modern basic requirements. This process remains with many drawbacks, as the machines need to be repaired and maintained, which is very expensive. Their creation requires huge expenditures because they are amazing machines [2, 3, 13].

5.2 Make people lazy

Excessive reliance on artificial intelligence reduces public awareness and users' desire to discover new aspects. Which makes the opportunity for developing the preserve of a few individuals, and this negatively affects the general perception and the main goal of making advanced software. People are often drawn to these innovations that may make people a problem in the future [3, 4].

5.3 Unemployment

Since AI replaces most of the monotonous tasks and various work with robots, the human impedance appears with less firmness making it capable of causing a significant problem within work processes. Each association hopes to replace the less qualified people with AI robots that may do work compared to greater productivity [3, 15].

5.4 No emotions

Of course, machines have greatly improved in terms of working efficiently, but they cannot replace the human bond that creates the group. They lack sensory and emotional perception, and devices cannot build a relationship with people which may be an essential asset when team management is involved [12, 16].

6 Conclusion

Computerized thinking, as well as innovation, is one aspect of existence that continually interests us and shocks us with ideas, topics, developments, new elements, etc. In the current year 2020, life has witnessed rapid and dramatic changes during the great spread of the COVID-19 epidemic, which was made by humans with artificial intelligence to limit the spread of the virus between countries. Artificial intelligence has used smart robots as an anxiety complex solution to reduce infection and as a magic solution to control world lives differently through online controls. Anyway, hidden tasks and so is the progress in modern organizations.

Towards the top, we scrolled through this examination through definitions of AI, brief history, uses of AI in public, and uses of AI. Uses of AI in the military, morals of AI, and three guidelines of mechanical autonomy. It is not the finish of AI, there's an entire other world to originate from it, who recognizes what the AI can accomplish for us presently, perhaps it'll be a whole society of robots.

References

1. Naudé, Wim. "Artificial Intelligence against COVID-19: An early review." (2020).
2. Shi, Feng, Jun Wang, Jun Shi, Ziyan Wu, Qian Wang, Zhenyu Tang, Kelei He, Yinghuan Shi, and Dinggang Shen. "Review of artificial intelligence techniques in imaging data acquisition, segmentation, and diagnosis for COVID-19." IEEE reviews in biomedical engineering 14 (2020): 4-15.
3. Tawafak, Ragad M., Awanis M. Romli, and Maryam Juma Alsinani. "Student assessment feedback effectiveness model for enhancing teaching method and developing academic performance." International Journal of Information and Communication Technology Education (IJICTE) 15, no. 3 (2019): 75-88.
4. Mohamadou, Youssoufa, Aminou Halidou, and Pascalin Tiam Kapen. "A review of mathematical modelling, artificial intelligence and datasets used in the study, prediction and management of COVID-19." Applied Intelligence 50, no. 11 (2020): 3913-3925.
5. Tatnall, Arthur. "Correction to Editorial for EAIT issue 2, 2020." Education and Information Technologies 25, no. 6 (2020): 5901-5910.
6. Malik, Sohail, Mostafa Al-Emran, Roy Mathew, Ragad Tawafak, and Ghaliya AlFarsi. "Comparison of E-Learning, M-Learning and Game-based Learning in Programming Education–A Gendered Analysis." International Journal of Emerging Technologies in Learning (iJET) 15, no. 15 (2020): 133-146.
7. Tawafak, Ragad M., Ghaliya Alfarsi, Awanis Romli, Jasiya Jabbar, Sohail Iqbal Malik, and Abir Alsideiri. "A Review Paper on Student-Graduate Advisory Expert system."

In 2020 International Conference on Computing and Information Technology (ICCIT-1441), pp. 1-5. IEEE, 2020.

8. Tawafak, Ragad, Ghaliya AlFarsi, Jasiya Jabbar, Sohail Iqbal Malik, Roy Mathew, Abir AlSidiri, Mohanaad Shakir, and Awanis Romli. "Impact of Technologies During COVID-19 Pandemic for Improving Behavior Intention to Use E-learning." (2021): 184-198.
9. Tawafak, Ragad M., Sohail Iqbal Malik, and Ghaliya Alfarsi. "Impact of Technologies During the COVID-19 Pandemic for Improving Behavioral Intention to Use E-Learning." International Journal of Information and Communication Technology Education (IJICTE) 17, no. 3 (2021): 137-150.
10. Iqbal Malik, Sohail, Roy Mathew, Ragad M. Tawafak, and Ghaliya Alfarsi. "A web-based model to enhance algorithmic thinking for novice programmers." E-Learning and Digital Media 18, no. 6 (2021): 616-633.
11. Malik, Sohail Iqbal, Roy Mathew, Abir Al-Sideiri, Jasiya Jabbar, Rim Al-Nuaimi, and Ragad M. Tawafak. "Enhancing problem-solving skills of novice programmers in an introductory programming course." Computer Applications in Engineering Education (2021).
12. Tawafak, Ragad M., Ghaliya Alfarsi, and Jasiya Jabbar. "Innovative Smart Phone Learning System for Graphical Systems within COVID-19 Pandemic." Contemporary Educational Technology 13, no. 3 (2021): ep306.
13. Tawafak, Ragad, Sohail Malik, Roy Mathew, Mohammed Ashfaque, Jasiya Jabbar, Maryam AlNuaimi, Abdalla ElDow, and Ghaliya Alfarsi. "A Combined Model for Continuous Intention to Use E-Learning System." (2021): 113-129.
14. Malik, Sohail Iqbal, Ragad M. Tawafak, and Mohanaad Shakir. "Aligning and Assessing Teaching Approaches With SOLO Taxonomy in a Computer Programming Course." International Journal of Information and Communication Technology Education (IJICTE) 17, no. 4 (2021): 1-15.
15. Nayal, Kirti, Rakesh Raut, Pragati Priyadarshinee, Balkrishna Eknath Narkhede, Yigit Kazancoglu, and Vaibhav Narwane. "Exploring the role of artificial intelligence in managing agricultural supply chain risk to counter the impacts of the COVID-19 pandemic." The International Journal of Logistics Management (2021).
16. Nogueira, Raul G., Jason M. Davies, Rishi Gupta, Ameer E. Hassan, Thomas Devlin, Diogo C. Haussen, Mahmoud H. Mohammaden et al. "Epidemiological surveillance of the impact of the COVID-19 pandemic on stroke care using artificial intelligence." Stroke 52, no. 5 (2021): 1682-1690.
17. Sharifi, Abbas, Mohsen Ahmadi, and Ali Ala. "The impact of artificial intelligence and digital style on industry and energy post-COVID-19 pandemic." Environmental Science and Pollution Research 28, no. 34 (2021): 46964-46984.
18. Sharifi, Abbas, Mohsen Ahmadi, and Ali Ala. "The impact of artificial intelligence and digital style on industry and energy post-COVID-19 pandemic." Environmental Science and Pollution Research 28, no. 34 (2021): 46964-46984.
19. Jabbar, Jasiya, Sohail Iqbal Malik, Ghaliya AlFarsi, and Ragad M. Tawafak. "The Impact of WhatsApp on Employees in Higher Education." In Recent Advances in Intelligent Systems and Smart Applications, pp. 639-651. Springer, Cham, 2021.
20. Al Farsi, Ghaliya, Azmi bin Mohd Yusof, Awanis Romli, Ragad M. Tawafak, Sohail Iqbal Malik, Jasiya Jabbar, and Mohd Ezanee Bin Rsuli. "A Review of Virtual Reality Applications in an Educational Domain." International Journal of Interactive Mobile Technologies 15, no. 22 (2021).

An Application of Heuristic and Meta Dendral Expert System

Ragad M Tawafak[1*], *Ghaliya* Alfarsi[1,2], and *Sohail Iqbal* Malik[1]

[1]Department of Information Technology, AlBuraimi University College, Buraimi, Oman
[2] College of Graduate Studies, Universiti Tenaga Nasional, Information Technology department, Malaysia

Abstract. A mobile application is the best way to reach the largest possible number of users, and it is a smart idea and dealing with it is very easy. In the Corona pandemic, many Mobile applications appeared that helped the community with the necessary precautions, Therefore, we designed an application on the phone to allow patients to book an appointment and choose the appropriate health and medical services, and it also allows for a PCR examination to detect Coronavirus, By making an appointment and going to the patient's home while taking preventive precautions from Corona. Derivation motor is constantly refreshed as new ends are drawn from each new sureness in the learning base which triggers additional rules, heuristics and standards in the inference engine. This paper is going to design and implement an application of expert system, it's first since forever achievement DENDRAL which turned into a stepping stone in the Artificial Intelligence field. The model shows the essential parts of AI and the major icons to do the assessment and treatment within the situation of corona virus. The reveals is this System can be a greatly important expansion of Artificial Intelligence.

1 Introduction

We all go to the hospital to consult the doctor on the way to receive treatment for the diseases that afflict us, bearing upon ourselves the extreme fatigue that we feel and the lack of time [1]. But with the spread of the Covid-19 pandemic, people are unable to meet their appointments and are unable to come to receive treatment in hospitals for fear of catching Covid-19 [2], especially the elderly and children, as they are more susceptible to being infected with the virus Covid-19 and also because of the weakness of their immune system [3]. Therefore, from here we began to design a medical care application during Covid-19, and our application is an electronic platform that seeks to provide all medical services with the best quality and easiest possible ways to process and have all of them healthy and useful as it reduces the percentage of people contracting the virus and exposing their lives to risks so that the user can consult the doctor in a smooth way, and the electronic platform also specializes in providing all services to patients on the health side in the best and easiest way

* Corresponding author: raghad@buc.edu.om

possible to maintain the safety of the patient without Putting his life at risk [4]. Figure 1 is an example of using mobile application as a treatment tool effective within COVID-19.

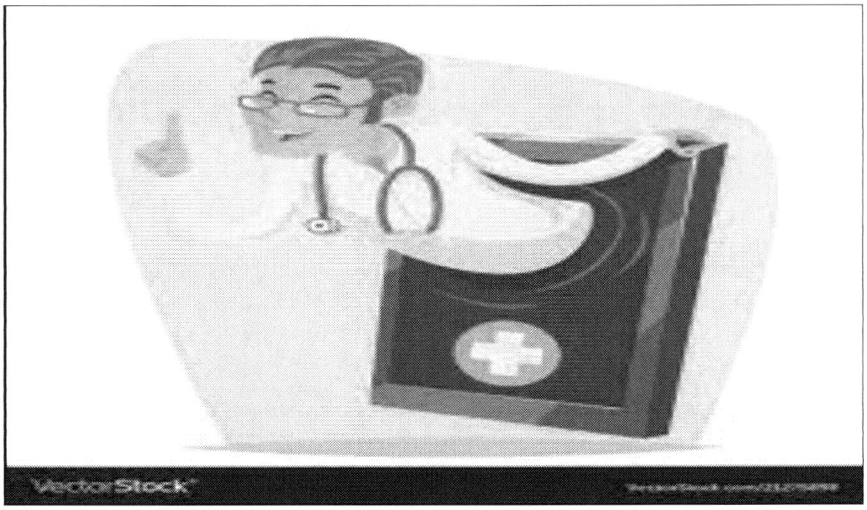

Fig. 1. Effective mobile treatment support

For that, an explicit errand in science was picked: help natural scientific experts in recognizing obscure natural atoms, by investigating their mass spectra and utilizing learning of chemistry. It was done at Stanford University by Edward Feigenbaum. By the pre-entered rule of guideline, The DENDRAL framework can naturally create atomic structures that can decipher unearthly information. The program is a first fruitful program that utilizes the learning of the issue itself as opposed to the Complex hunt innovation. The DENDRAL guides computerized reasoning specialists to understand that the keen conduct depends on impedance strategies as well as on the learning utilized in their Interference. Analysts start to construct the program that utilizes the tenets of the code to speak to the Knowledge to take care of the info issue. [10]
Numerous frameworks were gotten from Dendral, including MYCIN, MOLGEN, PROSPECTOR, XCON, and STEAMER. There are numerous different projects today to tackle the mass spectrometry reverse issue; however, they are never again depicted as 'man-made brainpower', similarly as structure searchers. The name Dendral is an abbreviation of the expression "Dendritic Algorithm". [11]

2 Background

Within computer science the DENDRAL Project. It was the main major enforcement of heuristic programming to test- tall analysis in an empirical science. It has carried out a high level of performance, ago it utilized a considerable measure of knowledge of chemistry. The DENDRAL programs were information driven. DEN- DRAL was the main rule-based system used to a "genuine world" issue. It has been used via, chemists. [10]

3 History of Dendral

The Dendral Project starts in 1965. Feigenbaum had been searching down an errand situation in which to exchange transaction of experimental acceptance (of models and speculations from information) and had arranged, his reasoning toward finding such an, engagement domain among the exercises that researchers do. Lederberg, a geneticist whose work in 1965 on exobiology included the mass spectra of ammunition acids, recommended the commitment of examining mass spectra—the development of estimation of natural atomic structure from mass ghostly information. Buchanan joined the exertion presently; his introduction was logic of science mixed with AI, a worry for the idea of logical revelation and the data forms essentiality. DENDRAL work was to a great extent test work. One of the soonest of the trial results was likewise maybe the most critical. That was the information is-control estimation, which has turned into the trademark by which numerous in AI recollect the DENDRAL Project. As we widened the breaking points of DENDRAL's capacities, what we found was that we required, more than everything else, more space explicit information of science and mass spectrometry (having all the more ground-breaking AI critical thinking strategies was valuable yet not urgent to our prosperity; more learning was essential). Toward this end, we join up the arranged stress of Djerassi, a world- class master in mass spectrometry, and with Djerassi his group of particularism, guests, and post-docs. One of our essential early inspirations, look the procedures of hypothesis development (in logical work, and somewhere else) was delayed for quite a long while. We settled on a choice (by and large right) to fulfill trial results and gain setting building background on an increasingly solid issue first: the theory arrangement issue of construing from one lot of otherworldly information one (or a couple) applicant sub-atomic structure(s). A fulfillment with DENDRAL drove us back to the first issue of hypothesis development, which currently showed up in a very explicit and solid frame that was "meta" to DENDRAL (henceforth the task's name, Meta-DENDRAL).[10]

Table 1. Expert System compound structure

Domain	**Organic chemistry- mass spectrometry**
Task	To distinguish molecular frame of strange complex from mass shadow data
Input	Histogram awarding mass digit
Output	attributes of structure of the compound
Architecture	plan-better -trial with constrained heuristic discussion

3.1 Case Study- DENDRAL

The Dendral venture is a genuine case of the rising development. Dendral was made at Stanford University to investigate chemicals. The endeavor was reinforced by NASA, in light of the way that an unmanned transport was to be dispatched to Mars and a task was required to choose the sub-nuclear structure of Martian soil, in light of the mass unearthly data gave by a mass spectrometer. The project being in 1965. Edward Feigenbaum, Bruce Buchanan and Joshua Lederberg encircled a gathering to deal with this testing issue. [10] Feigenbaum had been searching for a domain in which to look at methods of experimental acceptance, had organized his thinking toward finding such an errand circumstance among the activities that specialists do. Lederberg was a geneticist. He had worked in 1965 on exobiology incorporated the mass spectra of ammo acids, proposed the task of separating mass spectra. Afterward, Buchanan likewise joined the group and he presented the hypothesis of science blended with AI, and his advantage inclined towards the logical revelations and their related data forms. This venture was to a great extent an experimental one. Yet, it turned into a milestone for the Expert System field. [6]

Dendral represents Dendritic Algorithm. It is a procedure for thoroughly and non-repetitively specifying all the topologically distinct arrangements of any given set of atoms, according to the standards of compound valence. The focal issue of chemistry science is to choose the compound structure of atoms. The general issue which the DENDRAL programs apply is a basic, substantive issue in the science and that is structure explanation. Structure clarification is characterized as the procedure of assurance of the structure of a compound. The issue is basic in light of the way that the physical and in addition properties of the mixes are settled not just by quantities of the iotas, anyway by their topological and geometric plans moreover. A couple of observational methods are available for procuring information about the structure of a compound. Observable among these is mass spectrometry and DENDRAL at first kept an eye on the issues related just with this system, regardless of the way that it created to deal with the issues of structure explanation on progressively expansive terms.[10]

4 Types of Dendral

4.1 Heuristic Dendral

Heuristic Dendral is a program that utilizes mass vision or other exploratory information together with learning base of science, to create a lot of thinkable concoction structures that might be in charge of delivering the information. A mass area of a complex is created by a mass spectrometer, and is utilized to decide its sub-atomic weight, the whole of the majority of its nuclear constituents. For example, the compound water (H2O), has an atomic load of 18 since hydrogen has a mass of 1.01 and oxygen 16.00, and its mass range has a top at 18 units. Heuristic Dendral would utilize this info mass and the learning of nuclear mass numbers and valence rules, to decide the conceivable mixes of nuclear constituents whose mass would mean 18. As the load increments and the atoms turn out to be increasingly unpredictable, the quantity of conceivable mixes excess radically. Along these lines, a program that can diminish this number of hopeful arrangements through the procedure of theory development is basic. New chart theoretic calculations were imagined by Lederberg, Harold Brown, and others that produce all diagrams with a predefined set of hubs and association types (compound iotas and bonds) - with or without cycles. In addition, the group had the ability to demonstrate scientifically that the generator is finished, in that it delivers all diagrams with the predetermined hubs and edges, and that it is non-repetitive, in that the yield contains no comparable charts (e.g., perfect representations). The CONGEN program, as it became known, was developed largely by computational chemists Ray Carhart, Jim Nourse, and Dennis Smith. It was useful to chemists as a stand-alone program to generate chemical graphs showing a complete list of frame that satisfy the constraints specified by a user.[6]

4.2 Meta-Dendral

Meta-Dendral is machine learning frameworks that gets the order of conceivable synthetic structures and comparing mass spectra as info, and proposes a lot of standards of mass spectrometry that relate auxiliary highlights with procedures that create the mass spectrum. These principles would be supply back to Heuristic Dendral (in the arrangement and testing programs portrayed beneath) to test their suitability. In this way, "Heuristic Dendral is an execution framework and Meta-Dendral is a learning system". The program depends on two critical highlights: the arrangement produces test worldview and information building.

Plan-Generate-Test Paradigm: The arrangement produce test worldview is the sectional association of the critical thinking technique, and is a typical worldview applied by both Heuristic Dendral and Meta-Dendral systems. The generator (later named CONGEN) creates potential answers for a specific issue, which are then communicated as substance charts in Dendral notwithstanding, this is achievable just when the quantity of applicant arrangements is negligible. At the point when there are fundamental quantities of conceivable arrangements, Dendral needs to figure out how to put imperatives that discounts extensive arrangements of competitor solutions. This is the essential point of Dendral organizer, which is a "speculation development" program that utilizes "errand explicit learning to discover requirements for the generator". Last yet not slightest; the analyzer breaks down each proposed hopeful arrangement and discount of those that neglect to satisfy certain criteria. This component of plan produce test worldview is the thing that holds Dendral together. [21]

Knowledge Engineering: The primary point of learning designing is to accomplish a beneficial connection between the accessible information base and critical thinking techniques. This is conceivable through advancement of a strategy in which a lot of errand explicit data is encoded into heuristic projects. Thus, the principal fundamental segment of learning building is an extensive "information base." Dendral has explicit learning about the mass spectrometry system, a lot of data that frames the premise of science and diagram hypothesis, and data that may be useful in finding the arranging of a specific concoction frame illustration issue. This "information base" is utilized both to scan for conceivable compound structures that coordinate the info information, and to learn new "general guidelines" that assist prune seeks. The advantage Dendral gives the end client, even a non-master, is a limited arrangement of conceivable answers for check physically. [22].

5 Conclusion

Colossal commitment to the different territories has been made by the master framework from the most recent fifteen years. Master framework will keep on assuming an undeniably imperative job in the different fields. In this assessment, we have recognized and talked about the Expert Systems with its design and contextual fulfillment of its first achievement DENDRAL. Master System can be a greatly important expansion of Artificial Intelligence.

References

1. AlFarsi, Ghaliya., & ALSinani, M. Developing a Mobile Notification System for AlBuraimi University College Students. International Journal of Information Technology and Language Studies(IJITLS), 1(1). (2017).
2. Malik, S. I., Mathew, R., Tawafak, R. M., & Khan, I. GENDER DIFFERENCE IN PERCEIVING ALGORITHMIC THINKING IN AN INTRODUCTORY PROGRAMMING COURSE.
3. Tawafak, R. M., & Romli, A. B. Ruzaini bin Abdullah Arshah, and Sohail Iqbal Malik." Framework design of university communication model (UCOM) to enhance continuous intentions in teaching and e-learning process.". Education and Information Technologies, (2019).1-27.
4. Mathew, R., Malik, S. I., & Tawafak, R. M. Teaching Problem Solving Skills using an Educational Game in a Computer Programming Course. Informatics in Education, 18(2), (2019).359.

5. Clustering, A. S. International Arab Conference on Information Technology (ACIT). Higher Education, 165, 170. (2019)
6. Tatnall, A. Editorial for EAIT issue 2, 2020. Education and Information Technologies, 25(2), (2020). 647-657.
7. Tatnall, A. Editorial for EAIT issue 2, 2019. Education and Information Technologies, 24(2), (2019). 953-962.
8. Tawafak, R. M., Romli, A. B., Abdullah, R., Alfarsi, Ghaliya., Jabbar, J., Esbai, R., ... & Abdelhameed, W. Competitiveness & Quality.
9. Tawafak, Ragad M., Awanis M. Romli, and Maryam Juma Alsinani. "Student Assessment Feedback Effectiveness Model for Enhancing Teaching Method and Developing Academic Performance." International Journal of Information and Communication Technology Education (IJICTE) 15, no. 3 (2019): 75-88.
10. Tawafak, Ragad M., Awanis BT Romli, and Maryam Alsinani. "E-learning system of UCOM for improving student assessment feedback in Oman higher education." Education and Information Technologies 24, no. 2 (2019): 1311-1335.
11. Tawafak, Ragad M., Awanis BT Romli, and Ruzaini Bin Abdullah Arshah. "Continued Intention to use UCOM: Four factors for integrating with a technology acceptance model to moderate the Satisfaction of Learning." IEEE Access 6 (2018): 66481-66498.
12. Alfarsi, Ghaliya, Jasiya Jabbar, Ragad M. Tawafak, Sohail Iqbal Malik, Abir Alsidiri, and Maryam Alsinani. "Using Cisco Packet Tracer to simulate Smart Home."
13. Tawafak, Ragad M., Muamer N. Mohammed, Ruzaini Bin Abdullah Arshah, Mohanaad Shakir, and Vitaliy Mezhuyev. "Technology enhancement learning reflection on improving students' satisfaction in Omani universities." Advanced Science Letters 24, no. 10 (2018): 7751-7757.
14. Tawafak, Ragad M., A. B. Romli, Ruzaini bin Abdullah Arshah, and Rana A. Saeed Almaroof. "Assessing the impact of technology learning and assessment method on academic performance." EURASIA Journal of Mathematics, Science and Technology Education 14, no. 6 (2018): 2241-2254
15. Tawafak, Ragad M., Muamer N. Mohammed, Ruzaini bin Abdullah Arshah, and Awanis Romli. "Review on the effect of student learning outcome and teaching Technology in Omani's higher education Institution's academic accreditation process." In Proceedings of the 2018 7th International Conference on Software and Computer Applications, pp. 243-247. 2018.
16. Buchanan, B. G., & Duda, R. O. (1983). Principles of rule-based expert systems. In Advances in computers (Vol. 22, pp. 163-216). Elsevier.
17. O'Keefe, R. M., & O'Leary, D. E. (1993). Expert system verification and validation: a survey and tutorial. Artificial Intelligence Review, 7(1), 3-42.
18. Goita, Y., & Sidibe, M. (2021, March). Towards a Comprehensive Expert System for Coronavirus Disease. In 2021 7th International Conference on Information Management (ICIM) (pp. 18-23). IEEE.
19. Suppe, F. (2021). Artificial Intelligence and Computer Approaches to Clinical Medical. Logic of Discovery and Diagnosis in Medicine, 223.
20. Al Farsi G, Yusof AB, Romli A, Tawafak RM, Malik SI, Jabbar J, Bin Rsuli ME. A Review of Virtual Reality Applications in an Educational Domain. International Journal of Interactive Mobile Technologies. 2021 Dec 11;15(22).
21. Tawafak R, Malik S, Mathew R, Ashfaque M, Jabbar J, AlNuaimi M, ElDow A, Alfarsi G. A Combined Model for Continuous Intention to Use E-Learning System.

 https://doi.org/10.1051/itmconf/20224201009

22. Tawafak RM, Alsewari AA, Malik SI, AlFarsi G. A Tool Model Group (TMG) Development to Enhance Student Performance. In2021 International Conference on Intelligent Technology, System and Service for Internet of Everything (ITSS-IoE) 2021 Nov 1 (pp. 1-5). IEEE.

https://doi.org/10.1051/itmconf/20224201010

Privacy Preserving Data Mining Using Random Decision Tree Over Partition Data: Survey

Nashwan Adnan Othman[1*], *Mustafa Zuhaer Nayef* Al-Dabagh[1]

[1]College of Science, Knowledge University, Erbil, Kurdistan Region, Iraq

Abstract. The development of data mining with data protection and data utility can manage distributed data efficiently. This paper revisits the concepts and techniques of privacy-preserving Random Decision Tree (RDT). In existing systems, cryptography-based techniques are effective at managing distributed information. Privacy-preserving RDT handles distributed information efficiently. Privacy-preserving RDT gives better precision data mining while preserving information and reducing the calculation time. This paper deals with this headway in privacy-preserving data mining technology utilizing emphasized approach of RDT. RDT gives preferable productivity and information privacy than cryptographic technique. Various data mining tasks utilize RDT, like classification, relapse, ranking, and different classifications. Privacy-preserving RDT utilizes both randomization and the cryptographic method, giving information privacy for some decision tree-based learning tasks; this is an effective technique for data mining with privacy-preserving distributed information. Thus, in horizontal partitioning of the dataset, parties gather information for various entities but have data for all attributes. On the other hand, various associations may gather different data about a similar set of people. Thus, in vertically partitioned data, all parties gather data for the same collection of items. In all of these cases, both horizontal and vertical partitioning of datasets is somewhat inaccurate.

1 Introduction

Data Mining finds exciting data patterns, and insights from extensive databases. There are two phases in privacy-preserving data mining, the first is information collection, and the second is information publishing. In information collection, the holder stores information, which the proprietor gathers. In publishing, information could discharge to the receiver via the holder and the recipient mines published secured information. RDT algorithm builds numerous decision trees randomly. One essential structural part of RDTs is that they are entirely independent of the training information. Therefore, let us consider the two phases of the RDTs algorithm, training and classification. The training phase comprises building the trees and populating the nodes, which assumes the number of attributes is known based on the training dataset [1-3].

* Corresponding author: nashwan.adnan92@gmail.com

The procedure for producing a tree is as per the following. First, consider a features listing or note the attributes from the data set. Then, define a tree by randomly selecting a feature without utilizing any training information. The tree completes its development by reaching its uppermost limit. Then, it uses training data to refresh the statistics of each node. Note that only the leaf nodes need to record the number of cases of various classes grouped through the nodes in the tree. Finally, it scans the training information to upgrade the statistics in different random trees [1, 4].

RDT gives a better answer for the distributed data mining in concepts of privacy-preserving because of these reasons; random formation of the tree gives more security because to get prior information, one should find the entire classification model and cases. Since simple cryptographic strategy is slow and very little proficient concerning RDT because the branch of the tree is hidden to the outsider, the structure of RDT and its characteristics in which only the leaves of the tree are encrypted or decrypted [5, 6]. Four sorts of data mining tasks can utilize the same RDT code. RDT has another advantage because it can be made differentially private without losing data precision [7].

This paper proposes a proficient survey about privacy-preserving RDT. Its proficiency is its ability to maintain privacy and accuracy yet lessen computation time compared to existing algorithms. It uses an Iterative Dichotomiser 3 (ID3) and Boosting algorithm within an RDT, including a privacy-preserving algorithm. The algorithm developed safely builds an RDT for vertically and horizontally partitioned data. The Privacy-preserving RDT combines both randomization and cryptography techniques. This work executes a classification rule for data mining with privacy-preserving. Classification can be defined as storing information with similar features in the same class.

2 Random Decision Tree

2.1 Random Decision Trees Definition

A decision tree is a flowchart tree-like formation, which is utilized for decision analysis. In the decision tree, each internal node demonstrates the test of the attribute, the result of this test is represented by the branch of the tree, and the leaves demonstrate the class label or last result. The decision tree likewise demonstrates the alternative result for better comparison. A few classification algorithms for information discovery include neural networks, logistic regression, and decision trees. These decision tree classification techniques are variedly utilized with classification models, such as ID3, C4.5, CART, SLIQ, SPRINT, whereby every model heuristically employs splitting measures [6]. An RDT can be defined as the attribute selection done randomly since an arbitrary decision produces random information trees. RDT performs better than different models concerning calculation speed, in order to of the characteristics of random partitioning used as a section of tree development [8].

A RDT depends on two phases, training and classification, and the formation of a random tree is built entirely independent of the training information. While constructing every tree, first, begin with a list of attributes from the data set. Then, produce a tree by randomly choosing one of the attributes without utilizing any training information. Only the leaf nodes require to register the number of values of various subjects classified through the nodes in the tree. The training information is scanned exactly once to append the statistics in various random trees. While classifying a new example, the likelihood of various tree results is averaged to gauge a posteriori likelihood. The training stage consists of making the trees, defined as "BuildTreeStructure", and populating the nodes with the training example of information, defined as "UpdateStatistics". The assumption, all parties know the number of attributes depends on the training data set [1, 8].

In RDT, tree quits growing any deeper if a node becomes empty or there are no more instances of splitting in the current node, and the profundity of the tree exceeds a few limits. Random trees are utilized in communication networks to disseminate information from one node to another or gather information at a single assigned node.

2.2 Random Decision Trees Architecture

The structure of the privacy-preserving RDT appears below in fig 1. Distributed data acquisition depends on an issued query, which is a test dataset given as information tuples. That data alludes to obscure class names used to construct a privacy-preserving RDT. After ID3, the RDT classification happens, which utilizes a training dataset to delete noise and update the training dataset; then applies a random key on each classified part. Finally, the ID3 with RDT information produces a pattern dataset; again, used to classify the test dataset after boosting; this precisely reclassifies the training dataset information using a boosting algorithm. The verification and prediction phases classify information, foresee the class name, and prove the execution considering computational duration and accuracy [9-11].

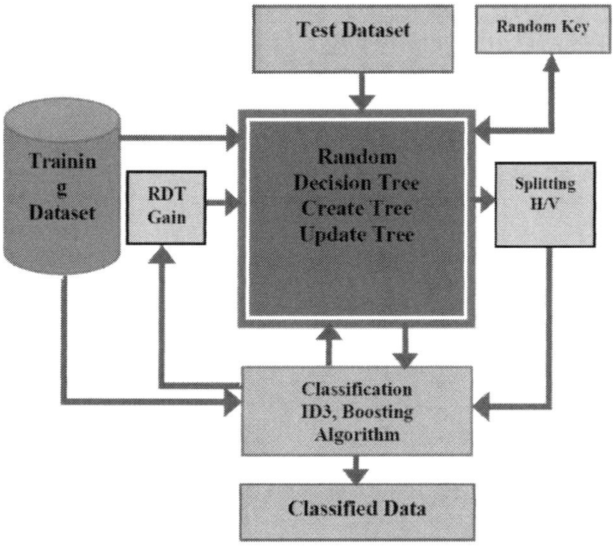

Fig. 1. Random Decision Tree Architecture.

3 Privacy-Preserving Data Mining

Privacy-preserving data classification aims to construct accurate classifiers without uncovering private data in the mined data. As of late, data mining has become noteworthy due to the proliferation of vast amounts of digital information [12]. Data mining includes analysing massive datasets, requiring computational procedures to ascertain patterns in those vast datasets. Furthermore, in the data mining process, data is shared between numerous clients while sharing information supplier needs to secure his sensitive information; hence, there is a method developed called privacy preserved data mining. Data mining privacy preservation is paramount today. Privacy preservation implies protecting gathered sensitive data gathered from various sources. The information patterns require collaborative discussions to refine handling methods employed in vast databases. However, these patterns can uncover sensitive data about individuals whose data alludes to the discussed patterns [13]. The idea of privacy-preserving data mining is to distinguish and forbid such disclosures, as evident in the sorts of patterns learned utilizing conventional data mining methods. In addition, privacy-preserving permitted the linkage of databases to associations by protecting privacy.

4 Partitioning of Data

4.1 Vertically Partition Data

Jaideep Vaidya et.al [14] introduce privacy-preserving decision trees over vertically partitioned data, summed up privacy-preserving variation of the ID3 algorithm for vertically partitioned data disseminated over at least two parties. With vertically partitioned data, every party will include similar entities and gather information for a heterogeneous collection of attributes [10]. Presently, the parties cannot autonomously make a random tree except if they disclose the attribute data; along these lines, all parties share essential attribute data (i.e., metadata). Presently, the parties can autonomously form random trees without sharing data. Subsequently, the parties must work together to make the random trees; these trees could be in a distributed shape [15]. As opposed to cases using horizontal partitioning, the formation of random trees does expose potentially sensitive data because the parties do not know the attributes owned by the alternate parties; this way facilitates a direct address of instances within entirely distributed trees.

4.2 Horizontally Partition Data

However, gathering and partitioning data from various entities for every attribute alludes to how to build and classify RDTs. Since each party shares the diagram, a straightforward solution is that they make a few random trees autonomously [16]. Nonetheless, every party can independently make the formation of the tree. All parties must cooperatively and safely calculate the parameters over the worldwide data set. Dissimilar to the fundamental RDT approach, there is no requirement to keep the class circulation at every non-leaf node; only some leaf nodes require this data. Presently, there are two possibilities. First, the tree formation is known to every member, and second, the formation of the tree is obscure to every member. The worldwide class distribution vector for every leaf node is known to none of the parties.

4.3 Data Partitioning Example

Let us consider a situation for predicting cars in a car exhibition; this requires defining a set of attributes, such as car name, model, colour, price, and quantity. For an example, these are the attributes:

Table 1. Distributed exhibition cars dataset.

Car Name	Model	Colour
Hyundai	2012	White
BMW	2003	Black
Volkswagen	2014	White
Kia	2009	Black
Audi	2012	White
Camry	2003	Black

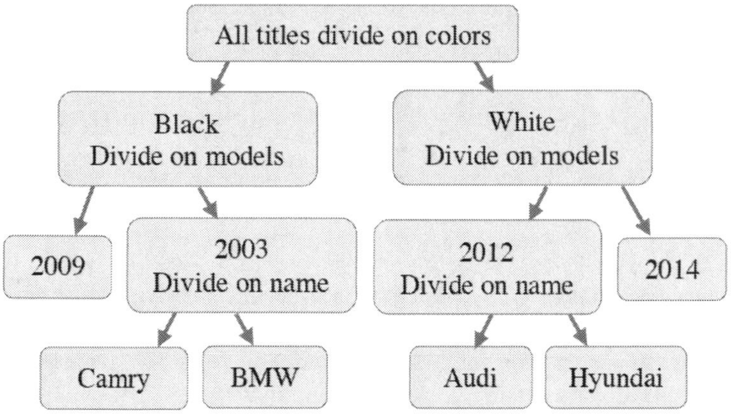

Fig. 2. A random decision tree with vertically partitioning data.

5 Privacy-preserving Random Decision Tree algorithm

The method outlined safely constructs RDTs for both horizontally and vertically partitioned datasets. Furthermore, the execution of the protocols provides a secure calculation at a low communication cost. Thus, RDTs can give excellent security with high profitability. Privacy-preserving RDT algorithm constructs different decision trees randomly. Building a tree is as follows: firstly, make a list of attributes utilizing the training dataset. Then, build a tree by randomly choosing listed attributes. Finally, the tree quits growing when it achieves the tree's limit of profundity. Thus, utilizing the ID3 makes a pattern dataset instead of a resulting classifier, which upgrades statistics at every leaf node utilizing the training dataset. To classify test datasets again requires the boosting algorithm with the help of a pattern. The algorithm builds every tree from the list of attributes and elects a "remaining" attribute randomly at every node. The consideration of an attribute as being "remaining" is if a similar attribute has not been selected in the past in a specific RDT node beginning from the root of the RDT to the present node. However, a continuous attribute can be chosen once in a similar decision path to the RDT. Thus, every time the class attribute is picked, a random threshold is chosen to enhance the conveyed of a distributed system for applications that work on a RDT overlay network.

6 Conclusion

Today, information privacy for various associations is paramount to expand their business since almost all organizations must share data without compromising privacy. This paper looks at randomization and cryptographic methods applied to sensitive information. The techniques outlined pertain to managing circulated information partitioned either horizontally or vertically across various locations within a secure framework of non-disclosure. This paper looks at the challenges of data mining tasks considering this backdrop. This paper concentrates on the technical plausibility of realizing privacy-preserving data mining. Privacy-preserving RDTs framework gives better precision with security and proficiency than RDT with privacy safeguarding system. The utilization of RDT can produce precise and sometimes better models with fewer costs. It improved the execution of RDT with a privacy safeguarding system. The approach utilizes distributed privacy-preserving RDTs, which influence a random structure to give complete privacy with less calculation than other methods. It can decrease computational time than RDT with a privacy-preserving system. Privacy-preserving RDT produces a highly accurate classifier, and learning is quick. This paper provides a case study of the horizontal and vertical partitioning of data using a privacy-

References

1. Hemlata B. Deorukhakar1, Prof. Pradnya Kasture2" Adaptive Random Decision Tree: A New Approach for Data Mining with Privacy-Preserving", Vol. 3, Issue 7, July 2015.
2. R. Agrawal and R. Srikant, "Privacy-Preserving Data Mining," Proc. ACM SIGMOD Conf. Management of Data, pp. 439 450, May 2000.
3. G. Jagannathan, K. Pillaipakkamnatt, and R.N. Wright, "A Practical Differentially Private Random Decision Tree Classifier," Proc. IEEE Int'l Conf. Data Mining Workshops (ICDMW '09), pp. 114-121,2009.
4. W. Du and Z. Zhan, "Building Decision Tree Classifier on Private Data,"Proc. IEEE Intl Conf. Data Mining Workshop on Privacy, Security and Data Mining, pp. 1-8, Dec. 2002.
5. G. Jaideep Vaidya, Senior Member, IEEE, Basit Shafiq, Member, IEEE, Wei Fan, Member, IEEE, Danish Mehmood, And David Lorenzi "A Random Decision Tree Framework for Privacy-Preserving Data Mining," Proc. IEEE Transactions On Dependable and Secure Computing, Vol. 11, No. 5, pp. 399-411, September/October 2014.
6. A. Dhurandhar and A. Dobra, "Probabilistic Characterization of Random Decision Trees," J. Machine Learning Research, vol. 9, pp. 2321-2348, 2008.
7. Jintu Ann John, Neethu Maria John,"Privacy-Preserving Random Decision Trees over Randomly Partitioned Dataset", vol.3, Issue 8, pp. 7746- 7750,2015.
8. W. Fan, H. Wang, P.S. Yu, and S. Ma, "Is Random Model Better? On Its Accuracy and Efficiency," Proc. Third IEEE Intl Conf. Data Mining (ICDM 03), pp. 51-58, 2003.
9. Matthew N. Anyanwu and Sajjan G. Shiva, "Comparative Analysis of Serial Decision Tree Classification Algorithms," International Journal of Computer Science and Security, (IJCSS) Volume (3): Issue (3).
10. Ming-Jun Xiao, Liu-Sheng Huang, Hong Shen and Yong-Long Luo, "Privacy-Preserving ID3 Algorithm over Horizontally Partitioned Data," Sixth International Conference on Parallel and Distributed Computing, Applications and Technologies (PDCAT'05).
11. Saeed Samet and Ali Miri, "Privacy-Preserving ID3 using Gini Index over Horizontally Partitioned Data," IEEE2008.
12. Priyank Jain, Neelam Pathak, Pratibha Tapashetti ,A.S. Umesh " Privacy-Preserving Processing of Data Decision Tree Based on Sample Selection and Singular Value Decomposition" In Proceedings the 9th International Conference on Information Assurance and Security, pp. 91- 95,2013.
13. J. Vaidya, C. Clifton, and M. Zhu, Privacy-Preserving Data Mining. Advances in Information Security first ed., vol. 19, Springer-Verlag, 2005.
14. J. Vaidya, C. Clifton, M. Kantarcioglu, and A.S. Patterson, "Privacy-Preserving Decision Trees Over Vertically Partitioned Data," ACM Trans. Knowledge Discovery from Data, vol. 2, no. 3, pp. 1-27, 2008.
15. Weiwei Fang and Bingru Yang, "Privacy-Preserving Decision Tree Learning Over Vertically Partitioned Data," International Conference on Computer Science and Software Engineering 2008.
16. M. Kantarcioglu and C. Clifton, "Privacy-Preserving Distributed Mining of Association Rules on Horizontally Partitioned Data," IEEE Trans. Knowledge and Data Eng., vol. 16, no. 9, pp. 1026-1037, Sept. 2004.

A New Adaptive Neuro-Fuzzy Inference System (ANFIS) Controller to Control the Power System equipped by Wind Turbine

Issam Griche[1,3*], *Sabir* Messalti[2,3], *Kamel* Saoudi[1] and *M.Yaakoub* Touafek[1]

[1] Department of Electrical Engineering, Faculty of Engineering, Bouira University, Bouira, Algeria.
[2] Department of Electrical Engineering, Faculty of technology, M'sila University, Algeria
[3] Mechatronics Laboratory (LMETR) - E1764200 Optics and Precision Mechanics Institute Ferhat Abbas University Setif 1 Algeria.

Abstract. This study proposes a new Adaptive Neuro-Fuzzy Inference System (ANFIS) controller to control the power system tuned by a wind turbine. The purpose of design is to improve the dynamical response of power systems after fault which the voltage controller has been improved. The effectiveness of the proposed approach is studied under different situations of three machines 9 bus power systems which the wind turbine is replaced by wind turbine equipped by ANFIS controller. The simulation results confirm that the tuning method is able to preserve optimal performances over wide range of disturbances. The results have demonstrated the high performances of the proposed technique in terms of low oscillation, ripple, rapidity and accuracy.

1 Introduction

Electrical energy is a flexible, adaptable form of energy, but it is difficult to store, and consumption by customers and coincidental demand are constantly varying. These requirements necessitate permanent transmission and provision of energy via a distribution system by high voltage for high powers and long distances and low voltage for medium and low powers and short distances. The system is regulated by means of requirements placed on producers, in particular the requirement to keep the current frequency at 50 Hz (or 60 Hz). Too much power results in an unwanted increase in frequency, while insufficient power causes a drop in frequency [1]-[2].

As the world shifts, its emphasis from fossil fuel to clean power, researchers are desperately searching for the perfect clean energy technology. Wind, solar, wave, tidal, biofuel and biomass are all experiencing renewed interest and new research funding. One alternative clean energy technology that bears looking at is wind powers which has been used for powering transportation and industrial equipment in cities and replace the power plants [3]-[4].

* corresponding author: griche_issam@yahoo.fr

Among them, high Large-scale integration of wind power generation has been tested. Previous studies indicate that the increase of wind power penetration in such isolated power systems in a reliable way, it is important to have good wind-power [5] for the correct planning and operation of the power system [6]. The focus of recent research has been on a stability problem during integration in power system [7]. However, very few publications can be found are available in the literature that discusses to solve this problem of stability [7].

However, a compromise between stability and accuracy is necessary for the performance of ideal controller. To surmount this requirement, Adaptive Neuro-Fuzzy Inference System (ANFIS) offers several attractive properties for the tracking control, such as lack of sensitivity to parameter variations, external disturbance rejection and fast dynamic responses [8].

In this study, a new ANFIS controller is proposed to control voltage of power system in presence of wind turbine which strong three-phase fault is applied.

The rest of this paper is organized as follows: In section 2 provides the model of synchronous generator and the wind turbine. Section 3 the design of proposed controller. Section 4 represents the simulation results. Finally, section 5 concludes the paper giving some comments and future directions.

2 Power system modelling

Let consider an N-bus power system model with m-machines. The standard two-axis model of the generator with field winding on the direct axis and damper winding on the quadrature axis, an IEEE type-I excitation system and a simplified turbine/governor model are adopted in this study.

2.1 Synchronous generator model

In this study, the dynamic behavior of machines is described by the following equations. For i=1...m,

$$\dot{\delta}_i = (\omega_i - \omega_b) \tag{1}$$

$$\dot{\omega}_i = \frac{\omega_b}{2H_i}\left[P_{mi} - P_{ei} - D_i(\omega_i - \omega_b)\right] \tag{2}$$

$$\dot{E}'_{qi} = \frac{1}{T'_{d0i}}\left[V_{exi} - E'_{qi} + (x_{di} - x'_{di})I'_{di}\right] \tag{3}$$

$$\dot{E}'_{di} = \frac{1}{T'_{q0i}}\left[-E'_{qi} - (x_{qi} - x'_{qi})I'_{qi}\right] \tag{4}$$

where $H_i, D_i, \omega_i, \delta_i$, P_{mi}, P_{ei} are respectively the inertia constant, the damping constant, the rotor angular velocity, the rotor angle, the mechanical and electrical input power of the i-th generator; ω_b is the rotating angular velocity. T'_{d0i}, T'_{q0i} are the d-axis and q-axis transient time constants; E'_{di}, E'_{qi} are d-axis and q-axis transient emf; x'_{di}, x'_{qi} are generator d-axis and q-axis transient reactances; I'_{d0i}, I'_{q0i} are the d-axis and q-axis of the generator currents, V_{ex} is the field voltage controlled by a voltage regulator.

2.2 Wind turbine model

The mechanical power P_T captured from a wind turbine of a blade radius (R) running in a wind stream of velocity V_w is given by the equation:

$$P_T = \frac{1}{2}\rho\pi R^2 C_p(\lambda,\beta)V_w \quad (5)$$

Where ρ is the air density, C_p is known as the power coefficient of the turbine. The tip speed ratio (TSR) is the ratio of turbine speed at the tip of a blade to the free stream wind speed and given by:

$$\lambda = \frac{R\omega_T}{V_w} \quad (6)$$

The power coefficient curve has been described in the literature by different fitted equations. In this paper, the power coefficient curve is approximated analytically by:

$$C_p(\lambda,\beta) = 0.5109\left(\frac{116}{X} - 0.4\beta - 5\right)\exp\left(-\frac{21}{X}\right) + 116\lambda \quad (7)$$

Where: $X = \dfrac{1}{\lambda + 0.08\beta} - \dfrac{0.035}{1+\beta^3}$

3 Design of the proposed controller

Adaptive Neuro-Fuzzy Inference System (ANFIS) is a combining between Neural Networks (NN) and Fuzzy Inference System (FIS), it can be applied to solve several optimization problems.

In this paper, the torque is optimized by the ANFIS controller shown in figure.1.

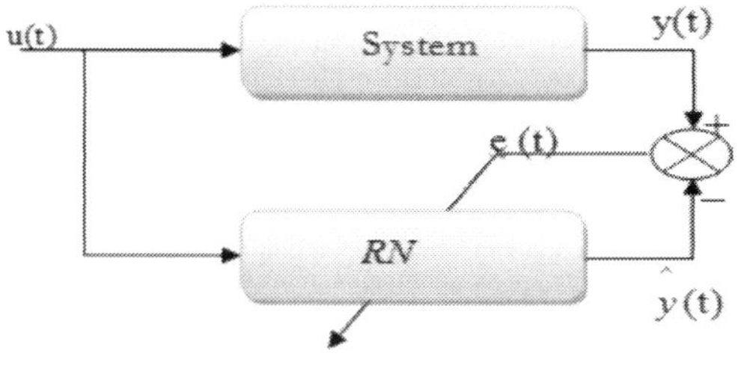

Fig 1. ANFIS controller proposed

The error equation $e_i = \hat{y} - y$ is used by an appropriate learning algorithm to adjust the parameters of the neural network. The Flow Chart of the proposed ANFIS controller is given by figure 2.

The fuzzy controller in our case contains 25 rules used to link inputs to the outputs. The Artificial neural networks (ANN) are used to adapt the membership functions and the determination of fuzzy rules. The configuration is Adaptive Neuro-Fuzzy Inference Systems (ANFIS). The method used for the Neural-fuzzy system adaptation is called hybrid: a combination of optimization by the least squares method and refined by the back propagation of the error shown in figure 3.

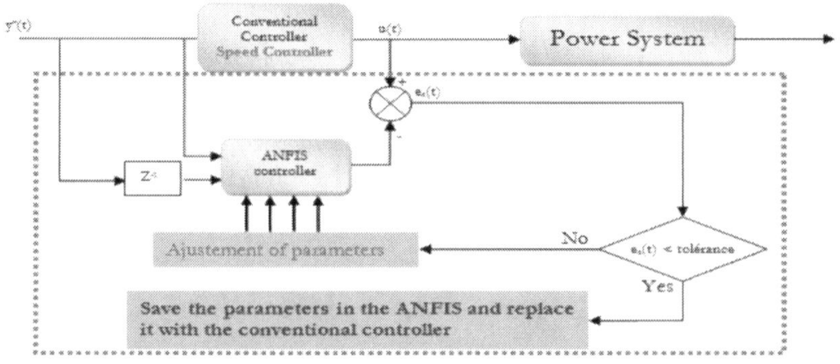

Fig 2. Flow Chart of the proposed ANFIS controller

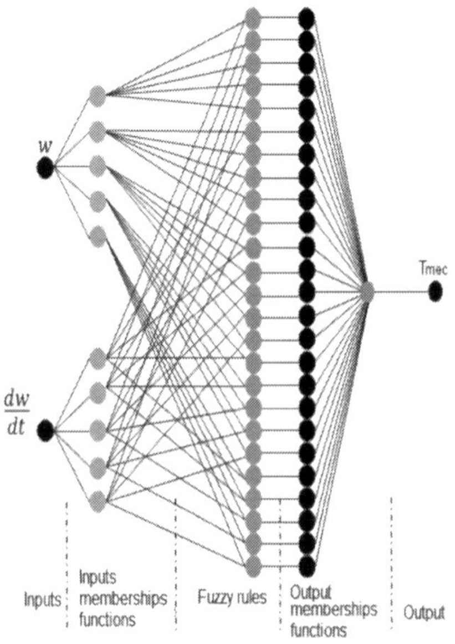

Fig 3. Proposed ANFIS controller

4 Simulations and discussions

In order to demonstrate the applicability, robustness and effectiveness of the proposed controller, simulation studies are carried out for the various fault disturbances and fault clearing sequences for three scenarios. The system being studied is simulated by MATLAB software. A 3 machines-9 bus IEEE system is used to evaluate the proposed model. Time domain simulations using Matlab-Simulink environment are performed for a strong three-phase fault of 0.2 sec applied to bus 4 at line 4-5.

4.1 Scenario without integration of wind turbine

In this scenario, the performance of the transient conditions is verified by applying a three phase fault at t = 0.2 s, on bus 4 at the end of line 4–5 is considered.

The oscillations are shown in Fig. 4. As seen in this figure, the oscillations are quickly damped out which means better dynamical response and faster damping of oscillations.

It can be noted that the system is transiently stable in these conditions because the

https://doi.org/10.1051/itmconf/20224201011

maximum value at which a fault is cleared after this disturbances.

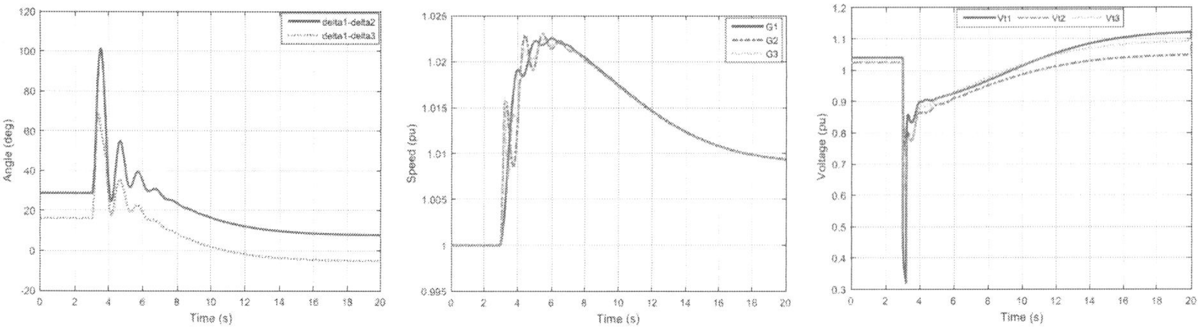

Fig.4. Simulation results without integration of wind turbine

4.2 Scenario with integration of wind turbine

In this scenario, another severe disturbance is considered for different loading conditions; that is, a three-phase fault is applied at the same above mentioned location in scenario 1.

For this simulation, it is assumed that a wind turbine occurred and replaced the generator 3. Fig. 5 displays the oscillations for the system response to this disturbance. It is obvious from this figure that the system loses its stability. As predicted, the criteria of stability in the presence of wind turbine have a longer value.

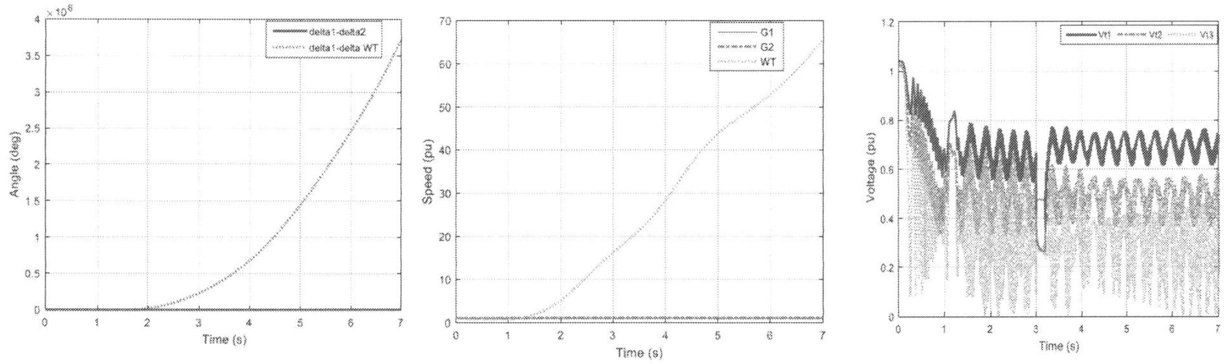

Fig.5. Simulation results with integration of wind turbine

4.3 Scenario with integration of wind turbine equipped by ANFIS Controller

Also, in this scenario, another severe disturbance is considered for different loading conditions; that is, a three-phase fault is applied at the same above mentioned location in scenario 1 and 2.

The fault is cleared with line tripping and the original system is restored upon the clearance of the fault. The system response to this disturbance is shown in Fig. 6.

Clearly, the ANFIS can suppress all oscillations satisfactorily. The best performance belongs to the proposed controller and has good damping oscillations and stabilizes the system quickly.

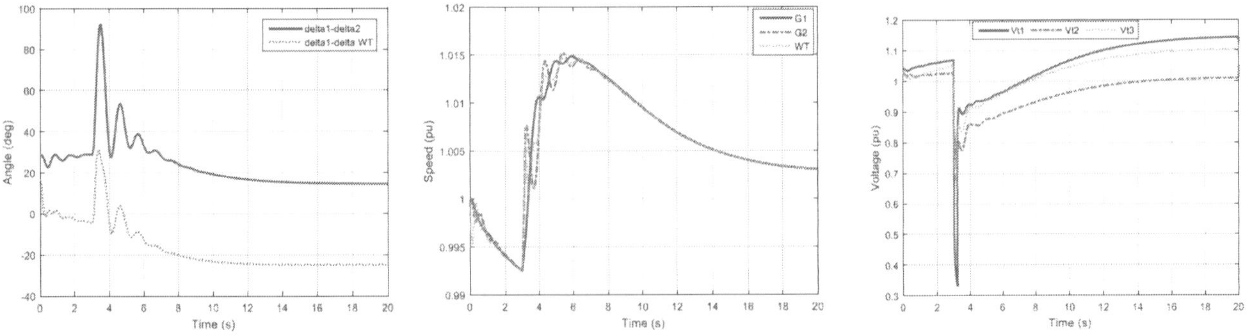

Fig.6. Simulation results with integration of wind turbine equipped by ANFIS controller

5 Conclusions

In this paper, a new ANFIS Controller is proposed to damp power system oscillations tuned by a wind turbine. To improve the dynamical response of power system, in the proposed approach, ANFIS controller is added to the control loop of wind turbine. These results demonstrate that the proposed controller confirms better disturbance rejection, keeps the control quality in the wider operating range, and damps the power system oscillations.

References

1. Speidel S, Braunl T (2016) Leaving the grid-The effect of combining home energy storage with renewable energy generation. Renew Sustainable Energy Rev, 60:1213–1224.
2. Hammons T.J (2008) Integrating renewable energy sources into European grids, Int.J. Electr. Power Energy Syst, 30: 462–475,
3. Bhaskar K and Singh SN (2012) AWNN-assisted wind power forecasting using feed-forward neural network. Sustainable Energy, IEEE Transactions on, 3(2): 306-315.
4. Li H.Z, Guo S, Li C.J and Sun J.Q (2013) A hybrid annual power load forecasting model based on generalized regression neural network with fruit fly optimization algorithm. Knowledge-Based Systems, 37: 378-387.
5. I.GRICHE, S. MESSALTI, K. SAOUDI, M. Y. TOUAFEK (2019) "A New Adaptive Neuro-Fuzzy Inference System (ANFIS) and PI Controller to Voltage Regulation of Power System Equipped by Wind Turbine", European Journal of Electrical Engineering (EJEE), Vol. 21, No. 2, pp 149-155.
6. I.GRICHE, S. MESSALTI, K. SAOUDI (2019) "Parallel Fuzzy Logic and PI Controller for Transient Stability and Voltage Regulation of Power System Including Wind Turbine",PRZEGLĄD ELEKTROTECHNICZNY, ISSN 0033-2097, R. 95 NR 9
7. I.GRICHE, S. MESSALTI, K. SAOUDI (2021) "Instantaneous Power Control Strategy for Voltage Improvement in Power Network Equipped by Wind Generator", Publication in Journal Européen des Systèmes automatisés JESA, Vol. 54, No.1, February, 2021, pp. 147-154.
8. I.GRICHE, S. MESSALTI, K. SAOUDI, M. Y. TOUAFEK, F.ZITOUNI (2021) "A New Controller for Voltage and Stability Improvement of Multi Machine Power System Tuned by Wind Turbine", Publication in Mathematical Modelling of Engineering Problems MMEP, Vol. 8 , No.1, February, 2021, pp. 81-88.

Design and Implementation of a Touchless Hand Sanitizer Dispenser Mobile Robot

Hind Z. Khaleel[1*], *Firas A.* Raheem[1], *and Mahdi* Azzam[1]

[1] Control and Systems Engineering Department, University of Technology – Iraq

Abstract. Two years ago, coronavirus COVID-19 spread all over the world. This virus causes the death of many people. The main problem is the way to maintain healthy hand wash in the circumstance of physical touch to the bottle that causes infection. The purpose of this paper is to design and implement a movable touchless hand sanitizer dispenser mobile robot that identifies the human body by processing the images taken from the camera to estimate the distance between the mobile robot and the human. A microcontroller board which is used to control four DC motors. The all design shows better activity than the usual and manual fixed touchless hand sanitizer design because the movable one can move with an acceptable error to the patient or a healthy person anywhere, he found. People can be protected from any virus or germs by using this proposed design. The main contribution of this work that our design system is useful in Anti-Covid-19 application and the total method has been collected the use of image processing, mobile robot, sensor hardware design, and the overall algorithm that control the application work.

1 Introduction

In récent years, technologies have been developed in all sectors of life. Humans developed robots to use in numerous applications of our life. One of the most important applications of the robot is the medical application. Hospitals use robots to make surgeries, besides, to carry medical equipment and for sterilization purposes, etc. [1]. A mobile robot design is useful in hand sanitizer operation, there are many studies presented in this field. The most priority has been given to hygiene as a prevention measure to avoid the transmission of the disease, as proposed by the World Health Organization (WHO), owing to the global severe dangerous respiratory syndrome coronavirus 2 (SARS COVID 2) epidemics Although the use of soap and water is the preferred way to clean one's hand, another option has recently become common. In their work, the researchers designed the AROGYAKAVACHAM device as an automatic sanitizer dispenser with a touchless temperature reader to protect people from SARS-Cov-2. Ultrasonic sensor, ordinal, L298 motor driver, IR sensor are attached to the device. Some drawbacks of their device: 1) the system requires to write temperature down of persons that entering the building even he or she sits six feet far from the device.

* Corresponding author: hind.z.khaleel@uotechnology.edu.iq

2) Through the lower solar insolation circumstances cause the solar panel of a system cannot to operate. 3) Temperature sensor did not give a correct reading while the human hands are not within the range and unsteady [2]. Hand sanitizers are represented as the products that are applied and cleaned over the surface of the hands to deactivate pathogenic microorganisms.These sanitizer products are made to dry quickly after the application are removing the need for soap, water, and drying helps for example the towels [3]. There is a severe assessment everywhere to control the COVID-19 disease and spread it to the world. All hospitals and the nurses have suffered from treatment the affected persons and stopping the virus from spreading to the neighbor persons. The sanitizer and the mask are providing to all over the place to protect the persons from spreading disease or virus and to kill the COVID-19 virus from the human hand. People's hands and mouth saliva are caused spreading of the virus in the world. The mouth spread is controlled by the usage of a mask and the hand wash sanitizer controls the human hand. The virus is also spread from one person to another, when each person's hands touch, use, and compress the same dispenser. It should be an automatic hand wash sanitizer dispenser, to control and maintain the spread from human to human. The researchers designed a fixed hand wash sanitizer. The motor has pumped the sanitizer liquid to the human by sensing the IR Sensor [4]. Robotics has appeared as one of the speedily industrialized engineering zones in the current-day invention and in numerous parts used in the industrial area to play out a varied task, which combines the industrial machines and parts like painting, and spot welding. Robots are predictable to remove an individual's association in a labor-serious and not used workplace. The authors design and develop a smart contactless hand sanitizer-dispensing with a mobile robot [5]. The main aim of this research can be briefly described as 1) In the previous papers [1 - 4], the authors designed the fixed hand sanitizer. The fixed system cannot reach the patient but the movable hand sanitizer reaches humans or patients at any location. 2) In this paper, a touchless hand sanitizer dispenser mobile robot has been designed and implemented using a Wi-Fi camera. The hardware camera is used to detect the human using image processing. Besides that, an ultrasonic sensor is linked to the mobile robot. It was useful for measuring the distance between mobile robots and humans. 3) In this work, the proposed hand sanitizer mobile robot is better than the hand sanitizer mobile robot as compared with [5]. In our system, the ultrasonic sensor with a camera computes the distance between the mobile robot and the human. In [5] the mobile robot moves randomly.

2 Modeling of a 4-wheel mobile robot

The kinematic model of the 4-wheel mobile robot consists of the following:

2.1 The kinematics of 4-wheel mobile robot

Kinematics can be defined as the mathematical study of the motion (position and velocity) without the forces consider [6]. The 2-DOF of differential drive 4-wheel mobile robot is shown in Figure 1. The velocity vector of the mass robot center V_G is perpendicular to the mass center axis as shown in Figure1.

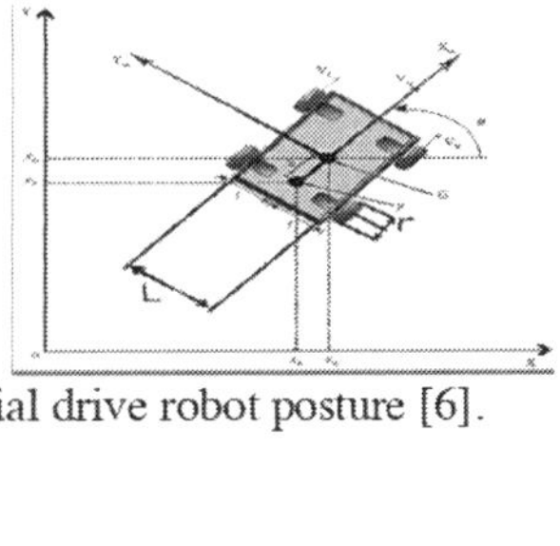

Fig. 1. Four-wheel mobile differential drive robot posture [6].

The x_G, y_G components and evaluated as [6]:

$$\dot{x}_G = V_G \cos(\theta) \quad (1)$$

$$\dot{y}_G = V_G \sin(\theta) \quad (2)$$

Where (x_G, y_G) are the coordinates of point G in the local frame, and θ is the heading angle of the mobile stand that is measured from the X-axis. By dividing (1) and (2), it became [6]:

$$\frac{\dot{y}_G}{\dot{x}_G} = \frac{V_G \sin(\theta)}{V_G \cos(\theta)} \quad (3)$$

$$\theta = \tan^{-1}\frac{\dot{y}_G}{\dot{x}_G} \quad (4)$$

The linear velocity of the wheel can be symbolized as (V), that having an angular velocity (φ), is stated as [6]; $V = \dot{\varphi}\, r$ (5)

The r letter is the radius of the mobile robot wheel. The velocity vector of the robot mass center is obtained by [6]: $V_G = \frac{V_L + V_R}{2}$ (6)

Due to (8) the linear velocity of the left and the right wheels are [6]:

$$V_L = \dot{\varphi}_L r \quad (7)$$

$$V_R = \dot{\varphi}_R r \quad (8)$$

r is the rotation of the mobile robot in the final humongous transformation. In addition, we substitute (7) and (8) in (6), we get [6]:

$$V_G = \frac{\dot{\varphi}_R r + \dot{\varphi}_L r}{2} = \frac{r}{2}(\dot{\varphi}_R + \dot{\varphi}_L) \quad (9)$$

$$\dot{\theta} = \omega = \frac{V_R - V_L}{L} = \frac{r}{L}(\dot{\varphi}_R - \dot{\varphi}_L) \quad (10)$$

Where $(\dot{\theta})$ represent the rotational velocity of mobile robot body.

With rewriting (6) and (10) the angular velocities for the left $(\dot{\varphi}_L)$ and right $\dot{\varphi}_R$ wheels are

$$\dot{\varphi}_R = V_G \frac{1}{r} + \frac{L}{2r}\dot{\theta} \quad (11)$$

$$\dot{\varphi}_L = V_G \frac{1}{r} - \frac{L}{2r}\dot{\theta} \quad (12)$$

L is the distance between the two wheels of a mobile robot. The kinematic matrix for the mobile robot is computed as [6]:

$$\dot{\mathbf{q}} = \begin{bmatrix} \dot{x}_G \\ \dot{y}_G \\ \dot{\theta} \end{bmatrix} = \begin{bmatrix} \cos(\theta) & 0 \\ \sin(\theta) & 0 \\ 0 & 1 \end{bmatrix} \begin{bmatrix} V_G \\ \omega \end{bmatrix} \quad (13)$$

$$\begin{bmatrix} V_G \\ \omega \end{bmatrix} = \begin{bmatrix} \frac{r}{2} & \frac{r}{2} \\ \frac{r}{L} & -\frac{r}{L} \end{bmatrix} \begin{bmatrix} \dot{\varphi}_R \\ \dot{\varphi}_L \end{bmatrix} \quad (14)$$

By combining (13) and (14), the equation is [6]:

$$\begin{bmatrix} \dot{x}_G \\ \dot{y}_G \\ \dot{\theta} \end{bmatrix} = \begin{bmatrix} \frac{r}{2}\cos(\theta) & \frac{r}{2}\cos(\theta) \\ \frac{r}{2}\sin(\theta) & \frac{r}{2}\sin(\theta) \\ \frac{r}{L} & -\frac{r}{L} \end{bmatrix} \begin{bmatrix} \dot{\varphi}_R \\ \dot{\varphi}_L \end{bmatrix} \quad (15)$$

3 Proposed touchless hand sanitizer dispenser of movable robot

The proposed flowchart of this work starts with the initialization of the values of the variables x, y, θ of a mobile robot as shown in Figure 2.

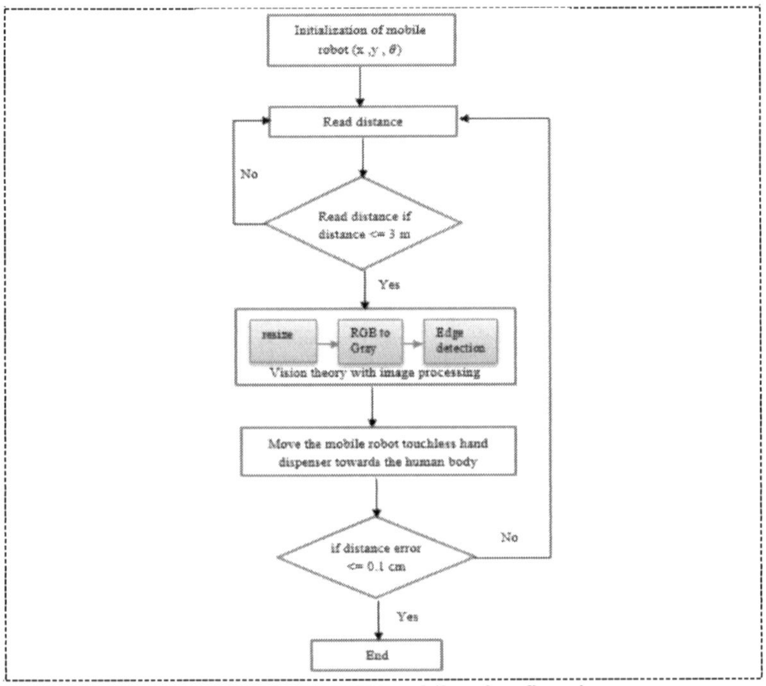

Fig. 2. Proposed touchless hand sanitizer dispenser mobile robot flowchart.

The calculation method was mentioned in the previous section. The mobile robot moves from the start point to the end (goal point). The ultrasonic sensor that is linked to the mobile robot is useful for measuring the distance between the mobile robot and the object (sensor) using sound waves. If the distance is ≤ 50 cm the movable robot stop and the vision (camera) captures the person's image. The personal image is processed using image processing methods in order to get the edge detection of the image as described below: The vision theory is based on image processing techniques. Image processing can be defined as a method to make many processes on the image. These processes are useful for enhancing and extracting useful information from an image [7]. In this work, the Wi-Fi vision device of the mobile robot is used for recording videos. Each video consists of multiple frames (images) of objects. The multiple images of the video are processed with three image-processing methods:

1) Resize image

In this method, an image resizes into (200 X 200) with the Bicubic interpolation approach. An interpolation method reduces the number of image pixels. The pixel B (r', c') is designed by interpolating the nearest 4 x 4 pixels where it starts with A (r, c) and finishes to A (r+2, c+2). Figure 3 shows the original image shape that has the two-scale factors symbolized as Sr and Sc of A. Sr: rows scale factor and Sc: column scale factor [8].

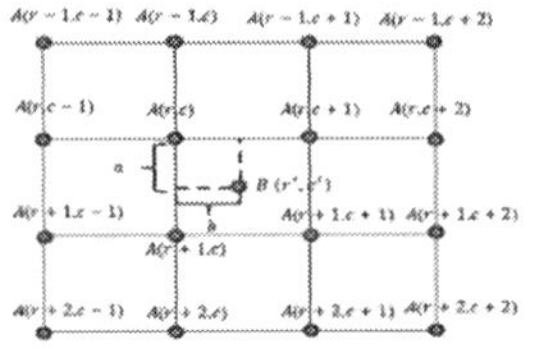

Fig. 3. Resize image with Bicubic interpolation method [8].

The new scale factors for image B are evaluated as (16) and (17) [8].

$$Sr = (R/R') \quad (16)$$

$$Sc = (C/C') \quad (17)$$

$B\ (r', c')$ calculated as

$$B(R',C') = \sum_{i=0}^{3} \sum_{j=0}^{3} a_{ij} A_{ij} \quad (18)$$

Where, A_{ij} illustrated as sixteen nearest of pixel values of B (R',C'). The factor a_{ij} is calculated by Lagrange (19) [2].

$$a_{ij} = a_i \mathrm{X}\, a_j \quad (19)$$

The two factors a_i and b_i are evaluated as (20) and (21) [2].

$$a_i = \prod_{k=0\ k\neq i}^{3} \frac{(r' - [Sr\times(x+k)])}{[Sr\times(x+i)] - [Sr\times(x+k)]} \quad (20)$$

$$b_i = \prod_{k=1\ k\neq j}^{3} \frac{(c' - [Sc\times(y+k)])}{[Sc\times(y+i)] - [Sc\times(y+k)]} \quad (21)$$

Where, a_i is the ith rows of A and b_i is the jth columns of A, and k is not equal to i. The x and y values for each row and column are divided with two scale factors Sr and Sc [8].

2) RGB to Gray image

In this section, the RGB (color) image is transformed to the gray image through the constant weights as in (22) [3].

$$Gray_{image} = 0.299 * R + 0.587 * G + 0.114 * B \quad (22)$$

The red, green, and blue are named as R,G, and B respectively. While 0.299, 0.587, and 0.114 represented the constant-coefficient weights. The summation of all constant weights is equal to one value [9].

3) Robert Edge Detection of image

Robert edge detection includes the extraction boundaries of the image. The image edges are extracted and converted from the grayscale image to the binary image [10]. The person image edges are extracted using the Robert Edge Detection method.

Using C++ and MATLAB programming languages, the main program is written in the Arduino Uno microcontroller that is attached to the mobile robot in order to move the mobile robot towards the human body with the all-hardware components described below. Then the position errors of e_x and e_y demonstrated as:

$$e_x = X_g - X_c \quad (23)$$

$$e_y = Y_g - Y_c \quad (24)$$

Where, $(X_g, Y_g)$is the goal point of the center body in the x-axis and y-axis of the mobile robot. (X_c, Y_c) is the current point of the center body in the x-axis and y-axis of the mobile robot. If position errors <= 0.1 cm the flowchart end and the mobile robot stop then the person can sanitize his hand with contactless. In this work, the hardware circuit of the mobile robot consists of an Arduino UNO Interfacing with the motor driver, ultrasonic sensor hc-sro4, two batteries 9 volt battery and ion lithium battery 3.7 volt supplied to the circuit. L298N motor drive is useful for controlling the 4 DC motors. An ultrasonic sensor measures the distance between the mobile robot and the person. The circuit of touchless hand sanitizer dispenser consists of an Arduino Uno linked with relay, IR sensor type tcrt5000, water pump type ultra-quiet submarine with 9 volt battery. The proposed design of the touchless hand sanitizer dispenser mobile robot is shown in the Figure 4 below.

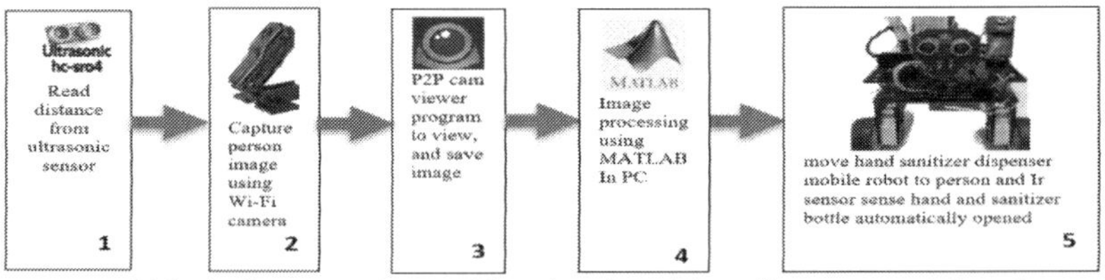

Fig. 4. The proposed block diagram of a touchless hand sanitizer dispenser mobile robot.

Figure 6 shows the proposed block diagram of this work. The first block includes the use of an ultrasonic sensor as mentioned in section 3 which is mechanically fixed on the mobile robot and can read the distance within 180 degrees in front of the mobile robot. The second block contains a Wi-Fi camera that takes one picture of the person. The third block includes a P2P cam viewer software installed on the PC and is used for connecting the Wi-Fi camera of the mobile robot with the PC. This software can view and save images on the PC. The fourth block contains the MATLAB program in order to process the image using image processing theory. In the last block, the mobile robot moves toward the person. The microcontroller card-based Arduino technique with H-bridge controls the four DC motors' direction of motion. When the mobile robot reaches the person, the IR sensor senses the human's hand, the microcontroller Arduino UNO sends a command to the relay, but it closes, and when the relay is closed, the water pump starts working and continues to work as long as the sensor senses an object. The water pump automatically raises the fluid from the bottom of the up using a relay. The fluid flows out and the human's hand can be sterilized without touching anything. The proposed system is enough to clean both human hands.

4 Experimental result

The human images captured from the Wifi-camera of the mobile robot are processed using image processing approaches. These methods are resizing, RGB to gray, and edge detection are mentioned in the above section according to equations (16), (17), (18), (19), (20), (21), and (22). Figure 5 shows image processing of four images captured for one person taken by the from Wifi-camera that fixed on the mobile robot during its motion towards the person and measured with three different distances (150, 53, and 10) cm as a case study for a test.

Fig. 5. Image processing of child that captured from mobile robot camera.

The movable robot moves from the start point (0,0) cm to the goal point (150,0) cm according to the ultrasonic sensor reading. The position errors of e_x and e_y demonstrated as in (23) and (24). The MATLAB results after 5 iterations are shown in Figure 6.

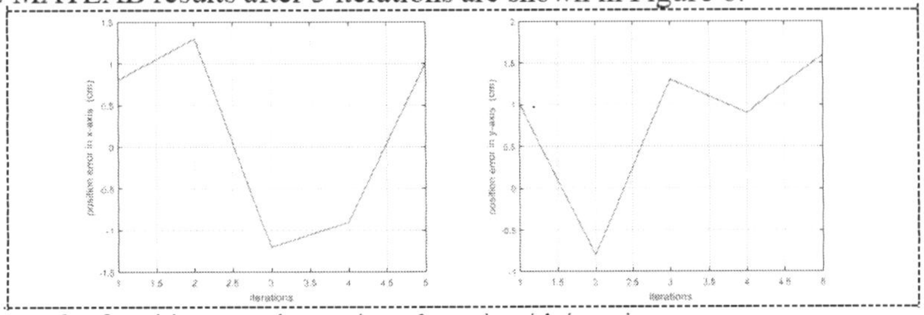

Fig. 6. The result of position error in x-axis and y-axis with iterations.

5 Conclusions

Touching dirty devices or tools can spread a lot of germs and viruses, as people with dirty hands also use them. A touchless hand sanitizer dispenser mobile robot has been designed to solve this problem because it offers a quick sanitizer way. With the proposed design of the touchless hand sanitizer dispenser mobile robot, there is no common contact point, which means less or no germs and viruses will be transferred from one person to another. The image processing technique is used to identify the person's body before moving the robot, and until reaching him by measuring the distance at each sample of motion. The approximate position error results are (1.3, -1.2) cm on the x-axis and (1.6, -0.8) cm on the y-axis. The experimental results of verifies the effectiveness of using the proposed design and the proposed technique of solution where the images are taken with less noise that enables the mobile robot to recognize the person's image. The main contribution of this work besides to be an extremely new Anti-Covid-19 application which is now in real need, is the total approach that include and collect the use of image processing, mobile robot, sensor hardware design, and the overall algorithm that control the application work.

References

1. Jeon S, Lee J. The simulator for performance analysis of multiple robots for hospital delivery. 2016 13th Int Conf Ubiquitous Robot Ambient Intell URAI 2016. Published online 2016:746-748. doi:10.1109/URAI.2016.7734105
2. Joshi HSV. AROGYAKAVACHAM—Automatic Hand Sanitizer Dispenser with Temperature Measurement. In: Proceedings of the International E-Conference on Intelligent Systems and Signal Processing. ; 2022:785-796. doi:10.1007/978-981-16-2123-9_61
3. Abuga K, Nyamweya N. Alcohol-Based Hand Sanitizers in COVID-19 Prevention: A Multidimensional Perspective. Pharmacy. 2021;9(1):64. doi:10.3390/pharmacy9010064
4. Srihari MM. Self-Activating Sanitizer with Battery Imposed System for Cleansing Hands. Proc 2nd Int Conf Inven Res Comput Appl ICIRCA 2020. Published online 2020:1102-1105. doi:10.1109/ICIRCA48905.2020.9183347
5. Eddy Mohammed, M. N. Y, Daoodd II, Siti Humairah Kamarul Al-Zubaidi, S. B, Omar Ismael Sairah, A. K. A-S. 2019 Novel Coronavirus Disease (Covid-19): Smart Contactless Hand Sanitizer-Dispensing System Using IoT Based Robotics Technology. Rev Argentina Clínica Psicológica. 2020;29(5):215. doi:10.24205/03276716.2020.1022
6. Said WK, Ahmed S, Kassim AB. Nonlinear Controller Design for a Mobile Manipulator Trajectory Tracking. IJCCCE. 2014;14(January):30-41.
7. Gonzalez RC, Woods RE. Digital Image Processing Third Edition Pearson. Pearson International Edition; 2008.
8. Hisham MB, Yaakob SN, Raof RAA, Nazren ABA, Wafi NM. An analysis of performance for commonly used interpolation method. Adv Sci Lett. 2017;23(6):5147-5150. doi:10.1166/asl.2017.7329
9. Hantoush Alrubaie SA, Hameed AH. Dynamic Weights Equations for Converting Grayscale Image to RGB Image. J Univ BABYLON Pure Appl Sci. 2018;26(8):122-129. doi:10.29196/jubpas.v26i8.1677
10. Selvakumar P, Hariganesh S. The performance analysis of edge detection algorithms for image processing. 2016 Int Conf Comput Technol Intell Data Eng ICCTIDE 2016. Published online 2016. doi:10.1109/ICCTIDE.2016.7725371

Natural Language Processing and Parallel Computing for Information Retrieval from Electronic Health Records

Ali Abu Salimeh [1], *Najah* Al-shanableh[2*], and *Mazen* Alzyoud[2]

[1] Hashemite University, Jordan,
[2] Al al-Bayt University, Jordan,

Abstract. In this paper, we review the literature to find suitable information retrieval techniques for EHealth. Also discussed NLP techniques that have been proved their capability to extract valuable information in unstructured data from EHR. One of the best NLP techniques used for searching free text is LSI, due to its capability of finding semantic terms and in rich the search results by finding the hidden relations between terms. LSI uses a mathematical model called SVD, which is not scalable for large amounts of data due to its complexity and exhausts the memory, and a review for recent applications of LSI was discussed.

1 Introduction

Information retrieval involves searching and extracting metadata from images, texts, and audio files [1]. Within the medical field, information retrieval is involved with the extraction of information related to the patient's records. In the modern-day context, medical information is obtained from the web, social media, hospital records, and journal articles [2]. The amount of data found in Electronic Health Records (EHRs) gives rich information about diseases and historical medical information that could be used to understand better and treat many medical cases [3]. Searching inside these records is a challenge to traditional information retrieval techniques due to the non-standard structure and free text in these records [4].
Many Natural language processing (NLP) techniques have been used to extract valuable information in unstructured data from EHR. One of the best NLP techniques used for searching free text is Latent Semantic Indexing (LSI) due to its ability to find semantic terms and rich the search results by finding the hidden relations between words [5]. Parallel computing is a type of computation where many calculations the execution of processes are carried out simultaneously in a single machine that has multiple processors execute multiple tasks simultaneously and have shared memory communicate with each other using a bus [7]. That will increase the system's accuracy because we do not lose any part of data relations. For Natural language processing from EHR, where massive computations happen, parallel computing can increase the accuracy while minimizing processing time [11].

* Corresponding author: najah2746@aabu.edu.jo

1.1 Electronic Health Records (EHR)

EHR contains information about a patient's medical history, treatment plans, medications, treatment dates, allergies, and laboratory and test results, among others [3]. The digital format of an EHR makes it possible to store and share the patient's information using networked information systems.
There are many challenges to understanding and repurposing EHR information for clinicians and researchers. Since a great part of the substance in the clinical record is in unstructured (free text) clinical documentation and it can collect any type of text character, utilization of natural language processing is frequently required. Regardless of these difficulties, scientists have been fruitful in duplicating known hereditary affiliations and making new revelations utilizing EHR information [4]. So Clinical concepts must be normalized if decision support and analytic applications are to operate reliably on heterogeneous EHR data.

1.1.1 Latent Semantic Indexing (LSI)

Latent semantic indexing is a technique used in information retrieval that helps in the analysis of documents to reveal the hidden relationships (latent) between the words or semantics to improve the understanding of the retrieved information [5]. LSI utilizes a technique known as singular value decomposition (SVD) to scan the unstructured data to understand the relationships between the concepts and terms included in the documents.

1.1.2 Apache Spark for EHR

Apache Spark is an open-source analytics engine that is used in machine learning and big data analytics. The analytics engine is very fast, which makes it suitable in parallel processing frameworks and large-scale big data analytics platforms. Spark is implemented using two main instances, notably a cluster management interface and a distributed storage system [6]. The ability to support data processing from multiple repositories makes Apache Spark an ideal solution for the development of information retrieval systems. Spark has many properties as Speed, Ease of Use, Generality, and Runs on Hadoop, Apache Mesos, Kubernetes, standalone, or in the cloud [7].

2 Information Retrieval and Electronic Health Records

Yousefu managed to design and develop a model that aimed at assisting the extraction and retrieval of EHR using scenario-based information. The model works by first creating a framework that defines the relationships between diseases, symptoms, and other related clinical information [8]. Popescu proposed a model that utilizes general word sequencing, which is based on the use of a dynamic algorithm and implemented using the fuzzy version of the Smith-Waterman model. The model uses the domain ontology to calculate and compute the word similarity matrix [9]. Traina and Rosa proposed a system that is used to retrieve the image-based content of electronic health records. The Similarity Retrieval of Images System (SRIS) is designed in such a way that it can perform a similarity query over images containing the patient's data [10].

2.1 Information Retrieval, Electronic Health Records and Natural Language Processing

Jain aimed at developing a framework designed to improve the paper-based information retrieval technique for accessing EHR. The proposed methodology utilizes a technique known as semantic query expansion and is aimed at supporting the retrieval of information related to the patient's current health condition [11].
MEDREADFAST is a hybrid browser with search capabilities that was proposed and described by Gubanov. The browser is supposed to simplify the process of accessing and retrieving medical information from big clinical data. MEDREADFAST was developed using a combination of several information retrieval techniques, including natural language processing, data management, and machine learning [12]. Pruski developed a framework that helps to address issues of encryption in the retrieval of EHR. The approach aimed at overcoming the barriers of encryption by combining the use of knowledge organizing systems and standard XDS metadata [13].
Yadav proposed the development of a personalized search engine that helps to refine the process of obtaining an internet-based patient's medical information [14]. Zhu managed to describe the development of a model aimed at improving information retrieval of EHR using discharge models. The framework involved the combination of various models that utilized natural language processing to assess the suitability of the information retrieval system [15].

2.2 Information Retrieval, Electronic Health Records, Natural Language Processing and Parallel Computing

Al-Qahtani proposed a novel approach that aimed at improving the process of information retrieval from e-health systems. The model uses Latent semantic indexing (LSI) algorithm and is implemented within The Health Improvement Network dataset (THIN). The main aim of the model is to improve the accuracy of the information retrieval process whereby the Term Document Matrix is utilized [16]. Hammad developed a model for the sole purpose of enhancing the storage and retrieval of XML-based electronic health records. The framework is aimed at improving the querying and indexing operations in an environment characterized by cluster computing [17]. Ufuk Parali also proposed a new algorithm reduced row echelon form IR method (rrefIR) with higher average similarity precision to Get more relevant and noise-free documents by applying (SVD) on the reduced row echelon form obtained by utilizing Gauss-Jordan method of the covariance of (TDM) [18].

2.3 Information Retrieval, Electronic Health Records, and Apache Spark

Md. Rezaul Karim and others propose a healthcare analytics framework with Apache Spark for real-time healthcare monitoring and related analytics by integrating heterogeneous data sources like the Internet of Things (IoT) enabled medical devices, Electronics Health Records (EHRs), picture archiving and communication system (PACS), wearable sensors, and public open health datasets. The ultimate goal of this research is to provide an effective, scalable, and secure framework for increasingly important public health surveillance [19]. One of the best NLP techniques used in previous research for searching free text is LSI due to its ability to find semantic terms and rich search results by finding the hidden relations between terms. LSI uses a mathematical model called SVD, which is not scalable for large amounts of data due to its complexity and memory exhaustion [18].
Many researchers address this problem and proposed solutions that are based on distributed computing witch it divides one task between several machines to achieve a common goal. Thus, memory systems are divided among the processors. Each machine can communicate

with others via the network, but this way affects the accuracy of the system because separating the data will lose part of its relations. Still, this way affects the system's accuracy because separating the data will lose part of its relations. An enhanced parallel information retrieval methodology for the LSI technique to analyse and processes unstructured data from EHR datasets in less time and with more accuracy is needed. Apache Spark, which applies parallel computing, can solve the LSI problem because it uses a very fast analytics engine that makes it suitable in parallel processing frameworks and large-scale big data analytics platforms.

3 Conclusion

In this research, we reviewed the literature on NLP and Parallel techniques that proved their capability to extract valuable information in unstructured data from EHR. Through previous studies, it appears that various techniques can be used to obtain information related to search operations within the free text. One of the best NLP techniques used for searching free text is LSI due to its capability of finding semantic terms and in rich the search results by finding the hidden relations between terms. Table 1 provide a summary of the related works.

Table 1. Summary of Related Works.

N	Reference	Proposed Framework or Algorithm	Goal	Testing Dataset	Major Results
1.	(Jain, Thao and Zhao, 2012)	Semantic Query Expansion (SQE)	Enhance EMR retrieval through SQE	Sample dataset with nursing notes from the community health center in a Midwestern metropolitan area	The proposed framework can improve the retrieval of EHR in a community health center.
2.	(Gubanov and Pyayt,2012)	MEDREADFAST a hybrid browser with search capabilities for big clinical data	Return the links of the most relevant webpages sorted by decreasing order.	10k natural language sentences composed of patient records from different clinical domains.	The results indicate that the hybrid browser yields more relevant results than keyword search while at the same time improving the user's experience.
3.	(Pruski and Wisniewski, 2012)	The framework utilizes two main approaches; the first one uses XDS metadata to describe the contents of CDA documents while the second approach involves the exploitation of XDS metadata content.	Address the issues of encryption in the retrieval of EHR.	Documents prepared using Clinical Data Architecture	The results indicate that the proposed framework improves the precision and recall of information retrieval and search processes.

4.	(Yadav and Poellabauer, 2012)	A new ranking algorithm combines both user query and health profile	Refine the process of obtaining an internet-based patient's medical information.	Personal Health Record system developed in-house	The search engine uses crawling and ranking algorithms to eliminate irrelevant results and ranks the top priority results respectively.
5.	(Traina, Rosa, and Traina, 2013)	Similarity Retrieval of Images System (SRIS) which uses two types of similarity: range queries and k-nearest neighbor.	access the patient's image data and then perform a similarity search to obtain similar results related to6the patient's data.	Image data sets obtained from a clinical hospital at the University of Sao Paulo at Ribeirao Preto-Brazil	The results indicating a high degree of accuracy in terms of the similarity results.
6.	Md. Rezaul Karim, Ratnesh Sahay, and Dietrich Rebholz-Schuhmann. (2015)	propose a healthcare analytics framework with Apache Spark for real-time healthcare monitoring and related analytics by integrating heterogeneous data sources	provide an effective, scalable, and secure framework for increasingly important public health surveillance	public open health datasets	SparkML is used to handle big static datasets coming from static data sources, Proposed Knowledge bases with Ontology Web Language (OWL2) and Open Biomedical Ontology (OBO) are used to generate rules, ontologies, and semantic annotation.
7.	(Al-Qahtani, Amira, and Ramzan, 2015)	Distributed Latent semantic indexing	Improve the accuracy of the information retrieval process whereby the Term Document Matrix is utilized.	THIN dataset	The proposed system was found to be highly effective, especially because it helped to eliminate invaluable information contained in the free text.
8.	(Reis, Bonacin, and Perciani, 2016)	Intention-based information retrieval for EHR	Investigate the means of considering intention in search engines	Dataset sample from Public hospital of "Aguas de Lindóia", São Paulo State, Brazil	The simulation results indicate that the intention-based technique is successful in a real-world medical environment.

9.	(Hammad and Banikhalaf, 2018)	Parallel Approach for Managing XML-based Electronic Medical Records	Improve the querying and indexing operations in an environment characterized by cluster computing.	Not mentioned	The results indicate that XML-based querying and indexing greatly improves the information storage and retrieval process in a distributed computing environment.
10	(Ufuk Parali, Metin Zontul, and Duygu Celik Ertugrul, 2019)	propose a new algorithm reduced row echelon form IR method (rrefIR) with higher average similarity precision	Get more relevant and noise-free documents by applying (SVD) on the reduced row echelon form obtained by utilizing the Gauss-Jordan method of the covariance of (TDM).	Not mentioned	The linear independence of basis vectors provided by the Gauss-Jordan operation makes the rrefIR algorithm retrieve more noise-free and relevant documents than LSI and COV algorithms.

References

1. C. Manningm, P. Raghavan, and H. Schütze, *Book Review: Introduction to Information Retrieval,* Natural Language Engineering, vol. **16**, no. 1, pp.100-103 (2010).
2. C. Smith, *Information retrieval in medicine: The electronic medical record as a new domain,* Proceedings of the American society for information science and technology, vol. **43**, no. 1, pp. 1-30 (2006).
3. K. Häyrinen, K. Saranto, and P. Nykänen, *Definition, structure, content, use and impacts of electronic health records: a review of the research literature,* International journal of medical informatics, vol. **77**, no. 5, pp. 291-304 (2008).
4. J. Denny, H. Xu, *Chapter 12-linking genomic and clinical data for discovery and personalized care,* methods in Biomedical Informatics, Sarkar IN, ed. Oxford: Academic Press, pp. 395-424 (2014).
5. S. Dumais, *Latent semantic analysis*, Annual review of information science and technology, vol. **38**, no. 1, pp.188-230 (2004).
6. J. Shanahan and L. Dai, *Large scale distributed data science using apache spark*, In Proceedings of the 21th ACM SIGKDD international conference on knowledge discovery and data mining, pp. 2323-2324 (2015).
7. A. Khumoyun, Y. Cui, and L. Hanku, *Spark based distributed deep learning framework for big data applications,* In 2016 International Conference on Information Science and Communications Technologies (ICISCT), IEEE, pp. 1-5, (2016).

 https://doi.org/10.1051/itmconf/20224201013

8. A. Yousefi, N. Mastouri, and K. Sartipi, *Scenario-oriented information extraction from electronic health records*, In 2009 22nd IEEE International Symposium on Computer-Based Medical Systems, IEEE, pp. 1-5 (2009).
9. M. Popescu, *An ontological fuzzy Smith-Waterman with applications to patient retrieval in Electronic Medical Records*, In International Conference on Fuzzy Systems, IEEE, pp. 1-6 (2010).
10. A. Traina, N. Rosa, and C. Traina, *Integrating images to patient electronic medical records through content-based retrieval techniques*, In 16th IEEE Symposium Computer-Based Medical Systems, Proceedings., IEEE, pp. 163-168 (2003).
11. H. Jain, C. Thao, and H. Zhao, *Enhancing electronic medical record retrieval through semantic query expansion*, Information systems and e-business management, vol.**10**, no. 2, pp. 165-181 (2012).
12. M. Gubanov and A. Pyayt, *MEDREADFAST: A structural information retrieval engine for big clinical text*, In 2012 IEEE 13th International Conference on Information Reuse & Integration (IRI), IEEE, pp. 371-376. (2012).
13. C. Pruski and F.Wisniewski, *fficient medical information retrieval in encrypted electronic health records,* In Quality of Life through Quality of Information, IOS Press, pp. 225-229 (2012).
14. N. Yadav, and C. Poellabauer, *An architecture for personalized health information retrieval*, In Proceedings of the 2012 international workshop on Smart health and wellbeing, pp. 41-48 (2012).
15. D. Zhu, S.Wu, J. Masanz, B. Carterette, and H. Liu, *Using Discharge Summaries to Improve Information Retrieval in Clinical Domain*, In CLEF (Working Notes), (2013).
16. M. Al-Qahtani, A. Amira, and N. Ramzan, *An efficient information retrieval technique for e-health systems*, In 2015 International Conference on Systems, Signals and Image Processing (IWSSIP), IEEE, pp. 257-260 (2015).
17. R. Hammad and M. Banikhalaf, *A parallel approach for managing XML-based electronic medical records*, In 2018 IEEE/ACS 15th International Conference on Computer Systems and Applications (AICCSA), IEEE, pp. 1-5 (2018).
18. U. Parali, M. Zontul, and D. Ertugrul, *Information Retrieval Using the Reduced Row Echelon Form of a Term-Document Matrix*, Journal of Internet Technology, **vol. 20**, no. 4, pp. 1037-1046 (2019).
19. M. Karim, R. Sahay, and D. Rebholz-Schuhmann, *A scalable, secure and realtime healthcare analytics framework with Apache Spark.*" In Proc. of the 2nd INSIGHT student conference on Data Analytics, The Insight Centre for Data Analytics, pp. 83-83 (2015).
20. M. Al-Qahtani, A. Amira, and N. Ramzan, *Enhancing the efficiency of information retrieval in e-health systems*, In 2015 British Computer Society Health Informatics Scotland Conference, (2015).

Software Defined Networking Automation Using OpenDaylight and Network Virtualization for security and scalability: a network enterprise case

Aws Jaff*

Pharmacy Department, Knowledge University, Erbil, Iraq

Abstract. Software Defined Network SDN is a technique that centralized the control plane in one device and the data plane (forward plane) in the other devices connected and communicated via SDN protocols, so these devices will get instructions from a central entity called a controller to manage the network efficiently. The controller generally has a global view of the network topologies and can broadcast any management policy to all devices in the network easily. SDN makes the recent network more responsive and flexibly managed by delivering interoperable and programmable network services to network organizations all around the world. In this work, two real critical scenarios were investigated against other scenarios to manifest the SDN ability to deal with them. Finally, results show that the opted implementation features are better than the others in term of testing parameters.

1 Introduction

It is all agreed traditional networks are too expensive to deploy and manage, too complicated to control, and too hard to change or update.[1] Software-Defined Networking (SDN) has emerged as a flexible programmable platform for managing and controlling networks. Thus, the main goal of SDN is to make the SDN network faster and more customized.[2]

In traditional networks, scalability and security are two of the most important obstacles when managing, configuring, maintaining the network. The huge number of appliances over big network enterprises make managing and controlling difficult since the manual configuration is needed to add, remove, update, or even maintain an actual appliance or the whole network or sub-networks [3]. Furthermore, a multi-vendor network requires network expertise to deal with network inconsistency, therefore extensive knowledge is required to achieve heterogeneous devices that complicate the network segmentation. On another hand, the time-consuming and error-prone will occur when deploying stage takes place by the network administrator which will be a highly administrative hassle to configure network standards. SDN overcomes the complexity caused by different network protocols and configuration interfaces that different vendors use for their devices. Another SDN advantage is cost-

* Aws Ahmed Jaff: aws.ahmed@knu.edu.iq

effectiveness since the use of virtualization can serve the SDN acquisition accomplishment of operational cost.[4]
SDN as a definition is the separation of the control plane and data plane of the network elements and devices. SDN is a logically centralized control unit with programable capability via Application Programming Interface (API). The separation of the two planes and their remote communication can be achieved through layered SDN architecture as Open Networking Foundation (ONF) that basically consists of three layers; 1. Device layer: enabled OpenFlow devices, 2. Control Layer: logical central controller, and 3. Application layer: the implementation applications that run on top of the controller.
Recently SDN received a lot of attention to solve the most problems of the traditional networks, the design of the SDN change the evolution of the networking by expressing network algorithms in a way that appropriate abstraction meets its specific application and without changing the underlying of the network infrastructure.[5]
SDN solve many challenges of traditional hardware networks; as simple and shouldn't be expensive, vendor-independency which easy to switch and upgrade among different vendors, and flexible; should be structured in current requirements as isolation, virtualization, and network optimization.[6] SDN also solves the legacy communication protocols in the traditional networking to be one logical centralized control unit that empowers the network automation as multiple logical control units. Finally, SDN gives a higher security level in managing network infrastructures and data centres with the ideal level of scalability for future network expansion. [5]

1.1 Literature review

The SDN implementation of networking is a new trend in today's technology industry, that's why there are few types of research works that are close to the field of our implementation work [7]. Many studies have been aimed to compare SDN architectures with their performance, but now those controllers were the foundation of nowadays controller such as Floodlight and OpenDaylight (ODL) [8]. Many aspects have been considered for building those controllers, based on a set of the requirements and these aspects were evaluated in the following areas: virtualization, Graphical User Interface GUI support, Interfaces, open sources, API, OpenFlow support, productivity, and modularity. Each controller was revealed to build those controllers each possess a critical weakness in their SDN performance [9].
Rationally, many advanced studies of performing widely used SDN-based OpenFlow controllers, doing an impractical analysis of the effectiveness of an average of workload, and maximum throughput on such controllers as Floodlight, and OpenDaylight. The variety has been witnessed in that studies regarding the analysis [10]. The ODL architecture and developing ODL model for network automation in a simple way for better understanding and intends of new controller innovation and accelerate the ODL adaptation for network programming [11]. The heavy relied on the Northbound (NB) and Southbound (SB) architectures, protocols plugins, and APIs communication that support OpenFlow 1.0 and 1.3 used for control, manage and monitoring the network including the orchestration services, Also, ODL supports clustering services where multiple controllers take place as one logical controller. The controller should not be limited with a southbound protocol such as OpenFlow and Northbound APIs by encouraging ODL adaptation ability of NB and SB of cluster redundancy, high availability, and scalability [12]. Some new methods illustrated in configuring and testing network devices by using automation for reducing configuration time, easier to maintain, and increasing the network stability [13].
Virtualization can be a single instance or combination of many such as Operating System (OS), networks, Application servers and storage devices. There are many types of virtualizations, but our concentration will be on three types of them: (i) Hardware, (ii)

Software, and (iii) network virtualization. Furthermore, virtualization provides numerous benefits such as cost-effective, efficient utilization of recourses, and fast accessibility [14]. Since the use of a Linux OS to run and build SDN devices and its infrastructure in this experiment. GNS3 is the main emulator to deal with that approach, GNS3 is an open sources multi-vendors emulator that provides virtualization environment that simulate real network devices. Moreover, the GNS3 market offers OpenvSwitch (OVS) to install in an easy way with full functionality support. OVS allows implementing policies such as permitting or denying flows between devices and users in network security, monitoring, quality of service QoS, and automated control [15].
OpenDaylight is open source SDN controller that supports multiple Southbound Protocols, as well as exposes open northbound APIs. OpenvSwitchs support both traditional and software-defined networking protocols such as OpenFlow (OF). In this topology, the commands have been used to write OpenFlow rules to OpenvSwitches that are running on GNS3.
OFM is an application that is used to visualize the OpenFlow topologies that run on OpenDaylight located on the northbound to manage the OpenFlow communication networks via RESTCONF API. RESTCONF API is an interface that used RESTCONF protocol, those interfaces are generated at the run time of YANG models. YANG models are Network Configuration Protocol (NETCONF) data modelling language used to model configuration and state data manipulated by the Network Configuration Protocol. YANG interface is dynamic generation, so not statically compiled java programming classes. an implementation has made to this interface on separate server than the same ODL server since our virtualized ODL can't handle multitasking. In another word, YANG model ran by OFM in our article as a visualizer tool to configure ODL controller [16].

2 Methodology and Implementation

In this work, an experimental test has been made on the ODL controller with topology shown in figure (1) with different case scenarios to show the SDN controller capabilities to deal with such scenarios. The SDN controller which control and manage the whole topology elements which are responsible to forward the traffic. Those switches known as Open Virtual Switches (OVS) can act as a real device. A test made with unlimited of (i) Round Trip Time (RTT) using ping command to check the connectivity among host and their linked switches via ICMP packets. (ii) Throughput and Burst Rate to check the maximum number of packets per second and per time-period. (iii) Response Time to measure how much of time needed to convey the packets and how much those packets may cost a time to travel completely from node to node. Then, Wireshark tool has been used to capture, monitor, and analyse the real time packets forwarding when the controller communicates with network devices. The subsection 2.1 to 2.4 show the implantation and scenarios. Finally, a comparison has been made with this work with other work regarding the mentioned parameters in the Results.

2.1 Network Emulator, Topologies and Networking Devices

In this experiment, the main network emulator is GNS3 version 2.2.27 to host those network devices. A star topology has been used to connect the network devises with class C network starts with IP address of 192.168.122.X/24 as depicted in figure (1). Then, a centralized Cisco switch installed to connect Windows server, ODL server, OFM server, with multiple OpenvSwitches that connect the Firefox end-user devices. The aim of Windows server in the topology to provide a DHCP IP pool to the network to prove that SDN controller can overcome the network scalability. The whole network connected to the internet with NAT technology to install the controllers from online resources. Finally, Wireshark tool has been installed to test and monitoring the network at microscopic level by capturing the network packets.

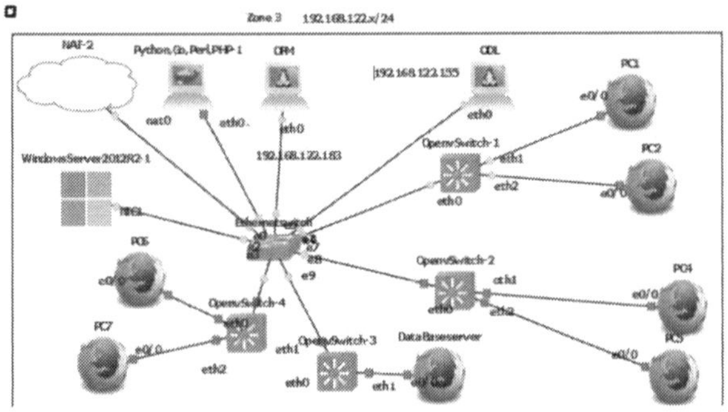

Fig. 1. The network topology

2.2 OpenDaylight controller (ODL) & OpenFlow Manager (OFM) Servers

In the topology, an OpenFlow management tool installed and used inside a separate Linux Ubuntu server (running on OFM server) to write OpenFlow rules to OpenvSwitches and visualize the OpenFlow topologies that run on OpenDaylight located on the northbound to manage the OpenFlow communication networks via RESTCONF API OpenFlow. Lastly, Open Daylight installed on the Open Virtual Appliance (OVA) running in VMware machine. ODL controller has been installed in another separate Linux Ubuntu server to provide the SDN core functionalities as shown in figure (1). In this experiment, the topology has four OpenvSwitch and two PCs are connected to each OpenvSwitch switch, and the OpenvSwitch switch is connected via Ethernet 0 to both ODL controller server, and OFM server. The first step was testing the connectivity among all devices in the topology. Then, a configuration has been made to all OpenvSwitch switches to connect ODL controller with IP address 192.168.122.155 on TCP port 6633. Once a connectivity achieved, the ODL sent a dump flow table 0 to all OpenvSwitch switches using OpenFlow 1.3 as shown in figure (3). This data link type is link layer discovery protocol (LLDP), a flow entry was written with a priority of a 100 and the traffic is being sent to the controller.

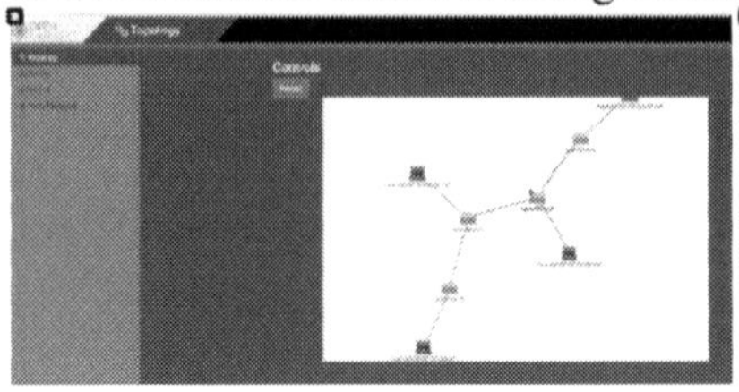

Fig. 2. OFM GUI

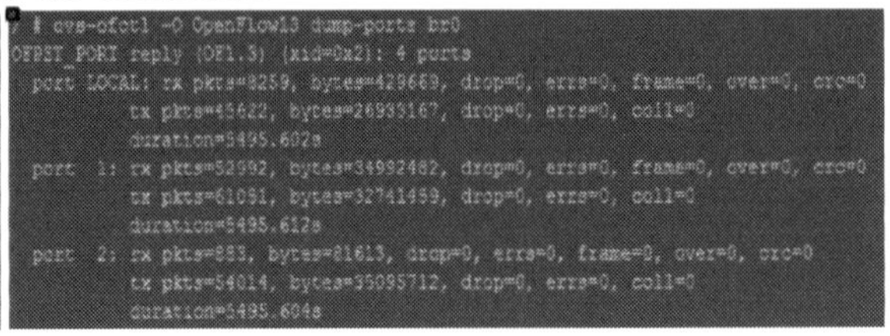

Fig. 3. Traffics on Port 2.

2.3 Case studies and scenarios

Case(i): For a core security reason, any PC must not access the ODL server for several reasons such as centre-directional administrative reason, server hacking, viruses, and malware affects/attack. So, the admin can block any network traffic coming from any non-authorized device to ODL server to have the accessibility and preserve this level of security. The admin can drop any packet coming from any PCs but still the PCs have the connectivity to all other devices except ODL server. Hence, from flow management menu the admin can add new flow entry to drop a any packet at any port. One ODL flow table can be assigned to configure hundreds OpenvSwitchs in case of future topology expandability, scalability, and network's adaptations. On the other hand, the same applicable to security reason, ODL offers the connectivity of any devices in the topology via MAC address as shown in figure 4. So, the case above can be also applied through MAC address restriction for thousands of unauthorized devices in one flow entry to reserve time and configuration effort.

Case(ii): Let's suppose that the network user in our topology heavily consuming the network traffic that effects the load of each network device with delivery latency and delivery failure or network device(s) failure (Load balancing). So, based on the network administration policy, it is better to limit the packet transfer speed to reduce the network overload among the enterprise branches as depicted in figure 4.

3 Results and Discussions

In this article, an investigation has been made to compare ODL controllers for an enterprise in a practical scenario. To compare and prove the controller capabilities, a Wireshark has been installed to analyse various packets. Since, OpenDaylight supports multiple southbound protocols, the investigation relied heavily on OpenFlow Switch Protocol's Read-State such as OFPT_STATS_REQUEST, OFPT_- STATS_REPLY, OFPT_PACKET_IN shown in figure 4. Those are to analyse the real-time packet's current configuration, statistics, and capabilities have been chosen randomly. Lastly, a comparison had been made with others ODL controller in [17] as shown in Table (1).

Table 1. ODL test parameters

Test Parameters	This article ODL controller test	Others ODL controller test
OpenFlow Response time	0.12065986 sec	0.190395-0.18951
OpenFlow Delay	0.009938 sec	0.444
OpenFlow length	20 bytes	20 bytes
OpenFlow configuration time	0.1596354 sec	0.212990-0.211245

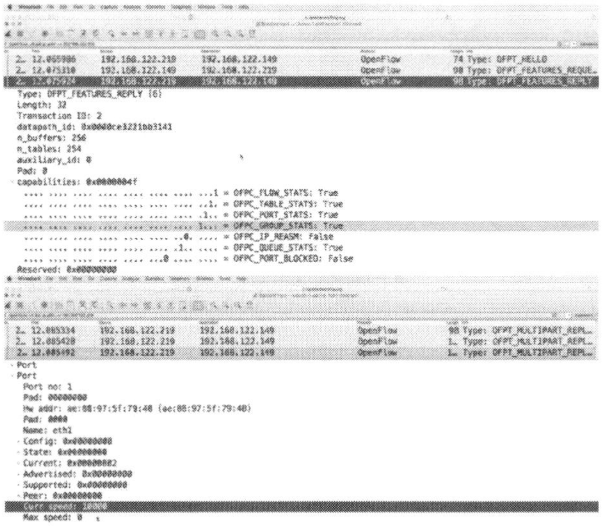

Fig. 4. Wireshark capturing test parameters

References

1. M. Mohammad & B. Eldin, A. & S. Mohamed. Software Defined Networking concepts and challenges. 79-90. 10.1109/ICCES.2016.7821979 (2016).
2. Casado, Martin & Koponen, Teemu & Shenker, Scott & Tootoonchian, Amin. Fabric: a retrospective on evolving SDN. 10.1145/2342441.2342459 (2012).
3. Zerwas, Johannes & Blenk, Andreas & Kellerer, Wolfgang. Optimization Models for Flexible and Adaptive SDN Network Virtualization Layers (2016).
4. Kreutz, Diego & Ramos, Fernando & Veríssimo, Paulo & Esteve Rothenberg, Christian & Azodolmolky, Siamak & Uhlig, Steve. Software-Defined Networking: A Comprehensive Survey. ArXiv e-prints. 103. 10.1109/JPROC.2014.2371999 (2014).
5. M. Casado, N. Foster, A. Guha 'Abstractions for software-defined networks', Magazine Communications of the ACM, Volume 57 Issue 10, October 2014 Pages 86-95. Available at http://www.cs.cornell.edu/~jnfoster/papers/sdn-abstractions.pdf (Accessed: 16 Jan 2020) (2014).
6. M. Casado, T. Koponen, S. Shenker, A. Tootoonchian 'Fabric: A Retrospective on Evolving SDN', HotSDN'12, August 13, 2012, Helsinki, Finland. Available at http://yuba.stanford.edu/~casado/fabric.pdf (Accessed: 15 Jun 2020) (2012).
7. Monaco, Matthew & Michel, Oliver & Keller, Eric. Applying Operating System Principles to SDN Controller Design. 10.1145/2535771.2535789 (2013).
8. Yeganeh, Soheil & Tootoonchian, Amin & Ganjali, Yashar. On scalability of software-defined networking. Communications Magazine, IEEE. 51. 136-141. 10.1109/MCOM.2013.6461198 (2013).
9. Zaalouk, Adel & Khondoker, Rahamatullah & Marx, Ronald & Bayarou, Kpatcha. OrchSec: An Orchestrator-Based Architecture For Enhancing Network-Security Using Network Monitoring And SDN Control Functions. Network Operations and Management Symposium (NOMS), 2014 IEEE. 10.1109/NOMS.2014.6838409 (2014).
10. Shalimov, Alexander & Zuikov, Dmitry & Zimarina, Daria & Pashkov, Vasily & Smeliansky, Ruslan. Advanced study of SDN/OpenFlow controllers. ACM Proc. CEE-SECR. 10.1145/2556610.2556621 (2013).
11. Rathi, Vipin & Singh, Karan. SDN Layer 2 Switch Simulation Using Mininet and OpenDayLight. 10.1007/978-981-10-8533-8_30 (2018).
12. Eftimie, Alexandra & Borcoci, Eugen. SDN controller implementation using OpenDaylight: experiments. 477-481. 10.1109/COMM48946.2020.9142044 (2020).
13. Wang, Lu & Sun, Meng & Tang, Shaoju. SCSCDaylight: Network Monitoring Tools for Software-Defined Networks Based on Opendaylight. 320-323. 10.1109/ICICAS48597.2019.00075 (2019).
14. Alaasam, Ameer & Radchenko, Gleb & Tchernykh, Andrei. Comparative Analysis of Virtualization Methods in Big Data Processing. 6. 48-79. 10.14529/js190107 (2019).
15. Kim, Wonho & Sharma, Puneet & Lee, Jeongkeun & Banerjee, Sujata & Tourrilhes. Automated and Scalable QoS Control for Network Convergence (2010).
16. L. Méndez, Jorge & Perdices, Daniel & Alvarez, Victor & G. Dios, Oscar & King, Daniel & Lee, Young. YANG data model for Flexi-Grid media-channels (2017).
17. Badotra, S., Panda, S.N. Evaluation and comparison of OpenDayLight and open networking operating system in software-defined networking. Cluster Comput 23, 1281–1291 (2020)

ITM Web of Conferences **42**, 01015 (2022) https://doi.org/10.1051/itmconf/20224201015
ICACS21

Simulation-based analysis of clock synchronization for underwater wireless sensor networks

Erik Diaz*, *Victor* Ramos, and *Rafael* Aguilar-Gonzalez

Universidad Autónoma Metropolitana, WiNetSys Team, 09340 Iztapalapa, Mexico City, Mexico

Abstract. Clock synchronization in underwater wireless sensor networks (UWSN) is a key procedure for the operation of this type of networks. Besides, due to the characteristics of the underwater channel, UWSNs face challenging constraints that make them different from other ad hoc wireless networks such as limited bandwidth, high error rates, high propagation delay, just to mention a few. Most of the existing literature on clock synchronization for UWSNs has been done through mathematical models. Thus, in this paper, we present a simulation-based approach to Light-Sync, which is a protocol synchronizing sensor nodes placed on the sea to a beacon on the surface. We use NS-2 along with AquaSim to study a UWSN with Light-Sync and prove its operation by means of discrete-event simulation. Moreover, our results are consistent with those found in Light-Sync and match accurately the synchronization accuracy regardless the number of nodes in the UWSN. Thus, we provide a simulation-based validation approach that makes more robust the proposal in Light-Sync.

1 Introduction

Underwater sensor networks are deployed to support a great variety of applications in the oceans such as temperature monitoring, disaster prevention, water salinity and water oxygen levels, to mention a few. Such applications are much more challenging to develop than traditional sensor networks since underwater communications present characteristics that make quite difficult the transmission in the channel. Traditional wireless sensor networks use radio or optical channels as the communications medium; thus, transmissions use the electromagnetic spectrum [1]. On the other hand, the downside of using the electromagnetic spectrum for UWSNs is that water greatly affects communications since it disperses the electromagnetic frequencies. Therefore, for underwater communications acoustic waves are preferred. Using acoustic waves makes communication possible in UWSNs, but this does not mean it is easy. In fact, several phenomena impact the quality of underwater communications. This is because the propagation speed of acoustic signals is about 1500 m/s, which clearly limits the maximum throughput of UWSNs.

Nowadays, communications in underwater networks are very useful for the range of applications cited above. An underwater wireless sensor network is commonly composed of

* Corresponding author: migueldyaz@xanum.uam.mx

a network of sensors deployed in the ocean sending data to a beacon placed at the surface. Figure 1 shows the general architecture for a UWSN. The sensor network relays data through multihop paths from the seabed to a station or a buoy, which acts as a supernode at the surface. Such a supernode may be equipped with an acoustic transducer, allowing multiple communications through surface towers and satellites.

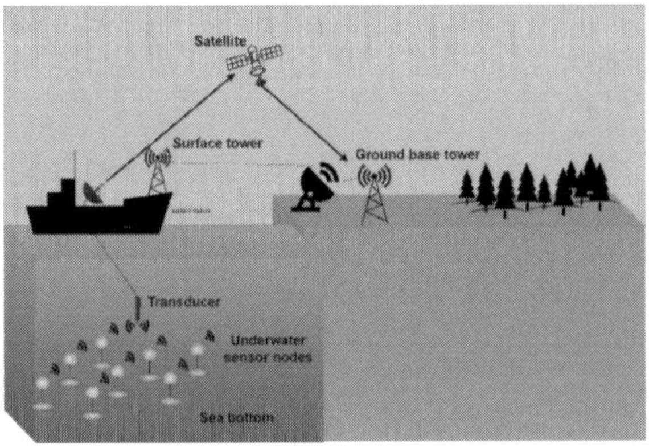

Fig. 1 General architecture of an underwater wireless sensor network (UWSN).

Often, for long time periods, the network topology is considered static, allowing the network to provide connectivity in this way. Nevertheless, such connectivity may change due to the small scale of ocean currents. With the aim of avoiding and reducing the impact of the characteristics of the acoustic channel, clock synchronization has been widely investigated in the last decade. Clocks in a UWSN lose synchronization mainly because of skew and drift. Most of the existing literature on clock synchronization for UWSNs considers the issues of high loss rates as well as the large propagation delays. The solutions provided are mostly mathematical models implemented as algorithms. On the other hand, an approach that has been less used is the use of a discrete-event simulator. Using a simulator has the advantage of considering the impact of a most complete set of phenomena rather than a mathematical model since in the latter several assumptions must be made to simplify analysis. Thus, in this paper by means of simulation with NS-2 and AquaSim, we evaluate the performance of Light-Sync, which is a clock synchronization protocol for UWSNs. In Section 2, we provide a brief review of the state-of-the art on synchronization in UWSNs; Section 3 describes the simulation of Light-Sync, a protocol to synchronize clocks in UWSNs. Section 4, presents the results obtained from validating Light-Sync by computing the time it takes the nodes to synchronize with respect to their distance to the buoy as well as the minimum-square error of the number of iterations. Finally, we sketch our conclusions in the last section.

2 State-of-the-art on synchronization in UWSNs

Work on synchronization for UWSNs mainly focuses on the low reliability of the underwater communications channel as well as on the characteristics of the network, i.e., a static or mobile topology. A series of exchanges of round-trip packets, followed by a linear regression is implemented to estimate clock skew and drift.

2.1 MU-Sync

The goal of MU-Sync is to compute clock skew and drift through two approximation phases [2]. The first one is called *skew and offset acquisition* phase, where sensor clocks are estimated with a two-step linear regression on a set of reference beacons, which are transmitted by a cluster that responds to the neighbor nodes. During the second phase, called *synchronization* phase, the cluster transmits clock skew and drift values of all neighbors. The approach followed by MU-Sync is by means of a cluster on a master-slave setup. However, the limiting factor in this protocol is the large message exchange between each node and the cluster, which requires several beacons, resulting in a message overload. Besides, it is assumed that the propagation delay is estimated with the average round-trip time, which causes additional errors.

2.2 D-Sync

In D-Sync, a cluster starts the synchronization process by transmitting a reference packet to which all the neighbors respond [3]. Through an estimation of the Doppler effect, the propagation delay is inferred by applying least squares estimators for the neighbor clock nodes with respect to the cluster, reducing in this way the estimation error. Like MU-Sync, D-Sync uses a master-slave approach; however, to get better efficiency in energy consumption, the approach is more complex which makes it even more sensitive to packet loss.

2.3 Mobi-Sync

Mobi-Sync considers networks composed of three different types of nodes: buoys at the surface equipped with GPS, supernodes operating with reference clocks always synchronized with the buoys, and low-complexity and low-cost nodes [4]. Mobi-Sync creates a spatial correlation between the mobile nodes to estimate the propagation delay using a model to infer the speed of ordinary nodes as a function of the speed of the supernodes. During the packet exchange phase, a node transmits a message to request clock synchronization. Each supernode neighbor schedules two response messages containing the speed vector and the timestamp of the MAC layer. The packet exchange phase ends when a node has collected enough data for a linear regression with a preset precision. After a fixed time interval, t_{r1}, each supernode transmits its first response packet trusting each supernode is located at a different distance with respect to the reference node to avoid collisions. In the interval t_{r2}, the supernodes transmit a second response packet. On the other hand, accounting with a more precise synchronization implies having a larger message exchange, which results in a higher energy consumption, which is undesirable.

2.4 Light-Sync

Light-Sync assumes the network is static since the nodes are anchored to the seabed. Thus, through a gateway it accurately gets the time by means of a GPS receiver [5]. Light-Sync focuses mainly on reducing energy consumption. It is executed in two phases, the first one consists of a series of full-duplex packet exchanges between a buoy P and a node A, which is the closest neighbor. The time between transmission and reception, ε, over P is fixed for node A. The first phase ends when node A has accurately estimated the clock parameters with a node B located within the communications range of node A, but not within that of P.

2.5 TSHL

TSHL (Time Synchronization for High Latency) is a clock synchronization algorithm for long-range and short-range networks [6]. It is executed in two phases; in the first one, each node computes clock bias through a linear regression from multiple beacons. During the second phase, clock drift is computed through a full-duplex packet exchange with a MAC timestamp. TSHL assumes a constant propagation delay, thus nodes must be stationary. The multihop network is formed of sensor nodes attached to buoys to platforms or boats.

3. Simulation of Light-Sync

As mentioned above, there is little work on evaluating clock synchronization protocols for UWSNs by means of simulation. Using a discrete-event simulator has the advantage of considering a wider range of phenomena at the physical and MAC layers since it makes no simplifying assumptions as mathematical models do. Moreover, providing a simulation-based approach validated with a mathematical model, provides a strongest approach for a protocol. Thus, in this paper we evaluate Light-Sync with the NS-2 simulator along with Aqua-Sim, which is an extension for UWSNs. Aqua-Sim allows to simulate the attenuation of the underwater acoustic channel, packet collisions caused by long delays, and also it is able to support three-dimensional simulation setups.

Table 1 lists the parameters used in the simulations of our scenario. We have executed 100 simulations for each initial position of the nodes. Thus, we deploy 30 underwater nodes randomly distributed not exceeding a volume of $100 \times 100 \times 100\ m^3$. Also, we put a buoy node at the center of the area placed at the surface (zero depth). During our simulations, each of the ordinary nodes is able to directly communicate with the buoy through a previous message exchange to synchronize, while the rest wait their turn to do the same operation

Table 1 Simulation parameters

Parameter	Value
Area	$100 \times 100\ m$
Depth	100 m
Transmission power	0.75 W
Receiving power	0.2 W
Propagation speed	1500 m/s
Routing protocol	VBF (Vector-based forwarding)
Simulation time	2000 s
Number of nodes	31
Maximum data packet size	40 bytes
Minimum data packet size	5 bytes
Coverage range	130 m

During the simulation, the propagation speed is kept constant. Each simulation campaign lasts 2000 seconds. After sending data, each node is able to pass to a waiting state saving energy in this way. The simulation scenario is illustrated in Figure 2.

https://doi.org/10.1051/itmconf/20224201015

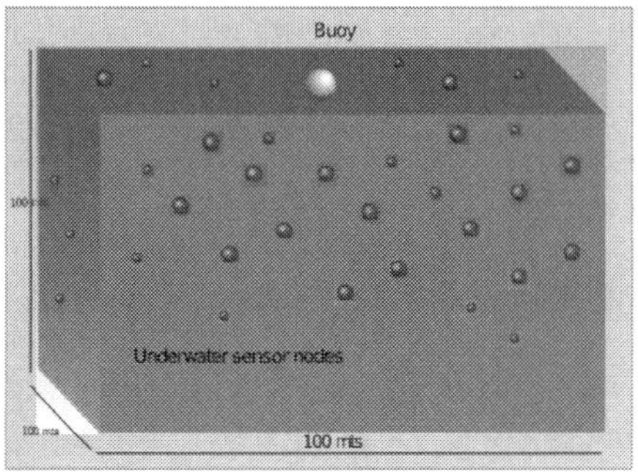

Fig. 2 Simulation setup for our Light-Sync evaluation.

4 Results

We execute campaigns of 200 simulations in order to get accurate results. The parameter of interest is the synchronization time of each of the 30 nodes distributed along the configurated region with respect to the buoy node at the surface. The synchronization procedure in Light-Sync starts when one of the nodes, mainly the closest to the surface, broadcasts a synchronization request message to the buoy, which contains a timestamp, $T_A(t_{i1})$. Then, at time t_{i2}, has received the message. Thus, after a time ε, the buoy responds to the node with a message carrying timestamps t_{i2} and t_{i3} of reception and transmission. To end this first part, the node receives such a message at time $T_A(t_{i4})$. Such message exchange between the node and the buoy is repeated I times until an accurate estimation of the clock skew and drift is obtained respect to the buoy.

Later on, the nodes that are furthest away from the buoy but inside its coverage range will repeat the same process. Then, the second phase of Light-Sync operates on the nodes outside the buoy coverage range but inside the coverage range of a node that has already estimated its clock values. Thus, we show in Figure 3 the linear regression we compute to confirm that the further the node is from the buoy, the largest its synchronization delay respect to it.

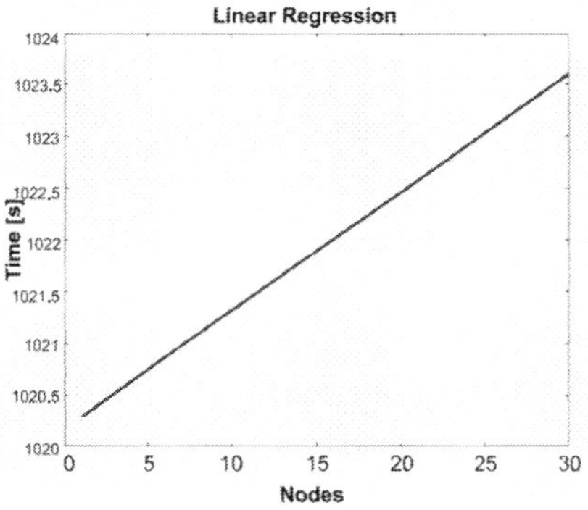

Fig. 3 Linear regression for clock synchronization in our scenario.

Finally, we compute the mean squared error (MSE) to analyze the exchange of messages between nodes and the buoy. Clock estimation is independent of the number of iterations, I, and the number of nodes in the network because each node is considered separately. Thus, Figure 4 shows the MSE computed from the obtained values during the communication between the buoy (P) and the closest node (A). We find that the estimators for clock skew and drift will be very efficient.

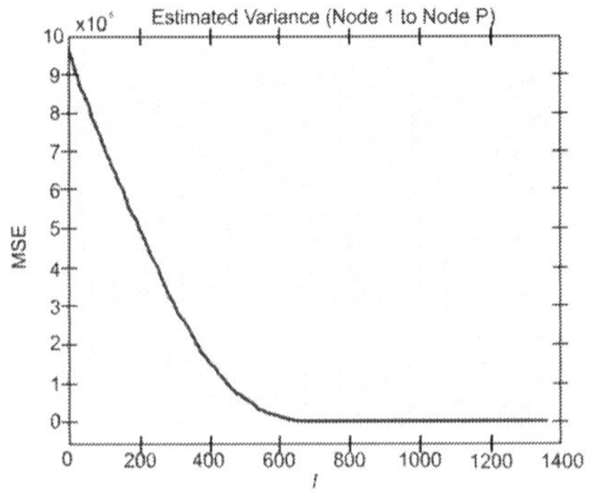

Fig. 4 MSE estimation versus the number of iterations, *I*.

5 Conclusions

We have evaluated, by means of simulation with NS-2 and Aqua-Sim the performance of Light-Sync, a well-known clock synchronization protocol for UWSNs. Most of the existing literature does not consider a simulation approach for evaluation. This is quite important, since such an approach does not take into account over simplifying assumptions often made by closed-form models. Moreover, providing a simulation-based evaluation offers a more complete and robust approach. Thus, we have validated the results presented in [5], which are consistent with those found with the analytical model provided. Our results match accurately those of Light-Sync about the synchronization accuracy, regardless of the number of nodes in the UWSN. The importance of validating synchronization algorithms by means of simulation is clearly important, as also shown in [7].

References

1. Fattah, S.; Gani, A.; Ahmedy, I.; Idris, M.Y.I.; Targio Hashem, I.A. A Survey on Underwater Wireless Sensor Networks: Requirements, Taxonomy, Recent Advances, and Open Research Challenges. Sensors (2020), 20, 5393. https://doi.org/10.3390/s20185393.
2. Chirdchoo, N., Soh, W.-S., & Chua, K. C. (2008). MU-Sync: A Time Synchronization Protocol for Underwater Mobile Networks. ACM International Workshop on Underwater Networks, 35–42. https://doi.org/10.1145/1410107.1410115
3. Lu, F., Mirza, D., & Schurgers, C. (2010). D-Sync: Doppler-Based Time Synchronization for Mobile Underwater Sensor Networks. ACM International Workshop on UnderWater Networks. https://doi.org/10.1145/1410107.141011510.1145/1868812.1868815
4. Liu, J., Zhou, Z., Peng, Z., Cui, J.-H., Zuba, M., & Fiondella, L. (2013). Mobi-Sync: Efficient Time Synchronization for Mobile Underwater Sensor Networks. IEEE ToPDS, 24(2), 406–416. https://doi.org/10.1145/1410107.141011510.1109/TPDS.2012.71
5. Zennaro, D., Tomasi, B., Vangelista, L., & Zorzi, M. (2012). Light-Sync: A low overhead synchronization algorithm for underwater acoustic networks, Oceans. https://doi.org/10.1109/OCEANS-Yeosu.2012.6263491
6. Syed, A. A., & Heidemann, J. (2006). Time Synchronization for High Latency Acoustic Networks. Proc. of the IEEE Infocom, 1–12. https://doi.org/10.1145/1410107.141011510.1109/INFOCOM.2006.161
7. Karim, S., Shaikh, F.K., Aurangzeb, K., Chowdry, B.S., Alhussein, M. Anchor nodes assisted cluster-based routing protocol for reliable data transfer in underwater sensor networks. IEEE Access (2021). https://doi.org/10.1109/ACCESS.2021.3063295

Novel Integration of Wi-Fi Signal and Magnetometer Sensor Measurements in Fingerprinting Technique for Indoors Smartphone positioning

Runahi Friad Qadr [1], *Halgurd* S. Maghdid [2*], *and Azhin* T. Sabir [2]

[1]MSc Student at Department of Software Engineering, Faculty of Engineering, Koya University, Kurdistan Region-F.R.Iraq. email: runahi.software@gmail.com
[2] Department of Software Engineering, Faculty of Engineering, Koya University, Kurdistan Region-F.R.Iraq. emails: first.last@koyauniversity.org.

Abstract. Smartphones are becoming more widespread, and location-based services (LBS) have become one of the most important uses in people's daily lives. While outdoor location is reasonably simple thanks to GNSS signals, however, indoor location is more problematic due to the lack of GNSS signals. As a result of the widespread deployment of alternative technologies such as wireless and sensors technologies, various studies on wireless-based indoor positioning have been conducted. However, each technology has its own limitations including multipath fading of wireless signals causes time-varying received signal strength as well as the accumulated error of the onboard sensors (i.e. sensor drift) resulting in poor localization accuracy. Motivated by these restrictions, this work integrates the applicability of two technologies for indoor positioning that are already available in smartphones by avoiding their limitation. The integration is based on fingerprinting-positioning technique by including magnetometer sensor measurements and WiFi signal strength. Android-based smartphones with low-cost sensors in real indoor scenarios are utilized to create a dataset and collect independent track tests to confirm results. The performance of different scenarios, such as Wi-Fi alone, magnetometer alone, and magnetometer-aided Wi-Fi, is compared. The experimental results show that the combination of magnetometer sensor and WiFi signal strength provides significant results in which leads to reducing the location error to 0.7224 meters.

1 Introduction

Utilizing smartphone applications to run daily life activities via users is increased day-by-day. For example, billing of foods/water/electricity, navigation in popular or unknown venues, tracking patient in hospitals, travel through airports, news and commercial advertisement, and sport/fitness activities are the most well-known applications on today's

* Corresponding author: halgurd.maghdid@koyauniversity.org

smartphones [1]. The unique feature of these applications is using location information. Further, indoor location information has acquired a huge attention as the demand for Location-Based Services (LBS) users via smartphones [2]. There has been a huge of number of techniques and technologies to run LBS applications. Pedestrian dead-reckoning with inertial sensors, fingerprinting with Wi-Fi, trilateration with beacon signals, and integration with high-sensitivity Global Navigation Satellite System (GNSS) receivers are only a few examples of prevalent techniques/technologies. Specifically, two types of positioning techniques based on Received Signal Strength (RSS) via WiFi technology can be implemented: 1) trilateration with a route loss model and 2) fingerprinting with a dataset. The first relies significantly on the signal propagation model and requires WiFi access point (WAP) locations. Fingerprinting, on the other hand, makes no such assumptions and uses a training dataset to calculate the likelihood of a user's location. In order to determine a user's location based on fingerprinting, various metrics can also be utilized to pattern-match between the online measurements and stored measurements in the dataset [3]. There are two phases in traditional fingerprinting which are offline and online phases [4]. The data collection method is carried out in an environment during the offline phase. The coordinates of RPs, as well as their respective signal intensities from Wi-Fi access points, are gathered. During the online phase, the indoor positioning algorithm estimates a location based on currently observed signal strengths and previously gathered information from RPs [5]. The received signal intensity can be modified by diffraction, reflection, scattering, and absorption during propagation in indoor situations, which is the main problem for positioning algorithms based on location fingerprinting, also because of the magnetic field's stability and distinctiveness, a number magnetic-based positioning techniques have been developed [6]. The abundance of WAPs in some interior areas, such as office buildings and shopping malls, allows the fingerprinting technology to function well [7]. As a result, Wi-Fi RSS-based fingerprinting location has grown popular, however the RSS measurements are not stable due to: 1) WiFi signal interference, WiFi chipset reading errors (i.e. ±4 dBm), and WiFi signal coverage in the vicinity [8]. Furthermore, with inertial sensors such as magnetometer sensor, the inside magnetic field anomalies induced by structural steel elements impact orientation estimate, although it is a signature for localization purposes in the meantime [9]. It is clear from the literature review that the majority of indoor positioning tests necessitate additional equipment installation [10-12] or they provide low positioning accuracy [13, 14]. To the best of our knowledge, so far, there is no a perfect solution for indoor localization while this work proposed an algorithm that combines features and advantages of RSS and magnetometer sensor together to accomplish or to provide better positioning accuracy at lower cost of survey of the area

Therefore, the key contribution of this work is: an integration of the benefits of combining magnetometer sensor measurements and Wi-Fi RSS values for positioning.

The rest sections of the paper are organized as the following, section 2 states the problem of fingerprinting positioning technique as well as investigate the current related solution. The proposed algorithm is deeply described in Section 3. Section 4 shows various trail experiments to demonstrate the validity of the proposed approach. Finally, section 5 summarizes the major accomplishments and outlines future plans.

2 The Proposed Algorithm

This study proposed a new integration of WiFi RSS values with magnetometer sensor measurements into fingerprint positioning techniques. The proposed approach this work is shown in Figure 1. The proposal via fingerprint technique for indoor positioning consists of two main phases which are offline and online phases. In the offline phase, the RSS values of WAPs and three axis magnetometer sensor measurements are collected in a site survey along the planned path. Further, the RSS values of WiFi access points and the magnetometer sensor measurements are used as positioning features which are gathered and stored with their relation location information (x and y coordinated) into the fingerprint dataset. An example of the gathered features during the survey is expressed in equation (1).

$$rec_{fea} = (x_{cor}, y_{cor}, RSS_{wap1}, RSS_{wap2}, \ldots\ldots RSS_{wapn}, Mag_x, Mag_y, Mag_z) \quad (1)$$

Where x_{cor} and y_{cor} are the location information at the fingerprint survey, $RSS_{wap1}, RSS_{wap2}, \ldots\ldots RSS_{wapn}$ are the RSS values of the WAPs signals in the vicinity, and finally the Mag_x, Mag_y, Mag_z are the three axis magnetometer measurements.

In online phase, the new RSS values of the WAPs and magnetometer measurements are then compared to the dataset's stored recorded values using the k-NN algorithm. Within k-NN, the comparison is applied by calculating the distance between the new RSS and magnetometer measurements and all previously stored or recorded data. The distance is determined in based on using the Euclidean distance, as shown in equation (2). Because of Android OS is free, powerful, and full, Android-based smartphone is utilized to collect the data records by enabling WiFi chipset and onboard magnetometer sensor

$$d_i = \sqrt{(ON\_RSS\_Mag - OF\_RSS\_Mag_i)^2} \qquad (2)$$

Where d_{ij} is the calculated distance between the online and offline values for the ith record, $OF\_RSSi\_Mag_i$ is the offline RSS and magnetic values at ith record, and ON_RSS_Mag is the online record values.

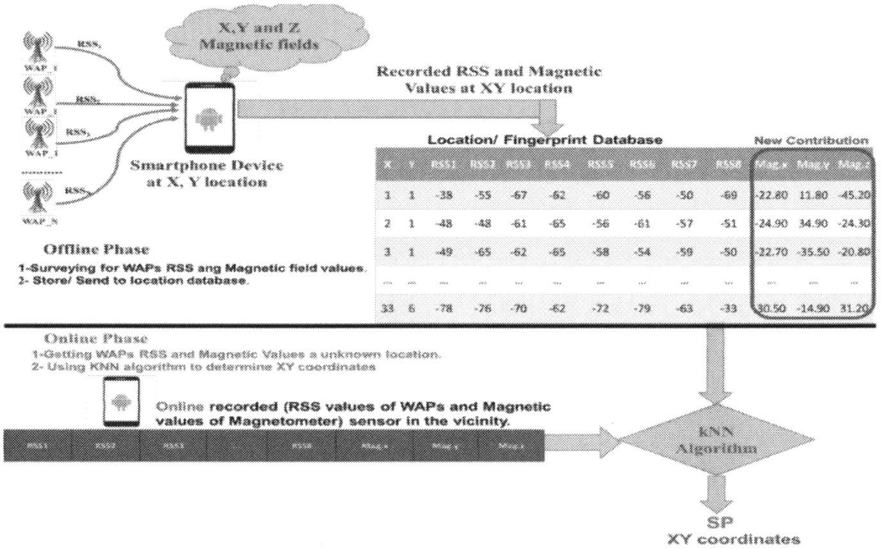

Fig. 1. The diagram of the proposed approach.

The distances for all the stored records are calculated, then in ascending order, the distances are sorted. Because of the real-time RSS values would be closed to some of the pre-stored RSS records, the first k distances are chosen. The k-value is determined via a fingerprint survey of the surrounding area or sites. Finally, for calculating smartphone (x, y) coordinates of the corresponding coordinates of the k-selected distances are used as the estimated smartphone location.

3 Dataset and Experimental Setup

To show the performance of the proposed approach, this study is tried to create a new dataset. The dataset is constructed via a survey at the first floor of the Koya Private Institute for Computer Science (Koya-PICS) building. The survey area is around 33 m by 6 m. At each meter square (at each fingerprint) the Wi-Fi RSS values and magnetometer measurements are collected. Thus, the number of the recorded data is around 198 samples. Further, due to instability of the WiFi RSS values and magnetometer values, this study attempts to get the values by three times at each fingerprint. Therefore, the total recorded data, in the offline phase, is equal to 594 records. Note the record sample is arranged as expressed in equation (1). During the survey, the RSS values of WAPs and magnetometer measurements are collected using an Android application on a smartphone.
To evaluate the proposed approach, MATLAB packages is used. A set of experimented is carried out. First, to provide truth fingerprint which has been used during the positioning accuracy, a MATLAB-simulated graph based on real survey of the building is shown in figure 2. The WAPs RSS in the vicinity, magnetometer measurements, and x, y coordinate values of the fingerprints are then stored. Next, as an initial experiment, the WAPs RSS values are depicted in figure 3. This is to show how the values of WAPs RSS values are changed along the path of the survey. As shown, there are eight WAPs RSS values.

Fig. 3. A simulated-MATLAB graph of the real survey for the Koya-PICS building.

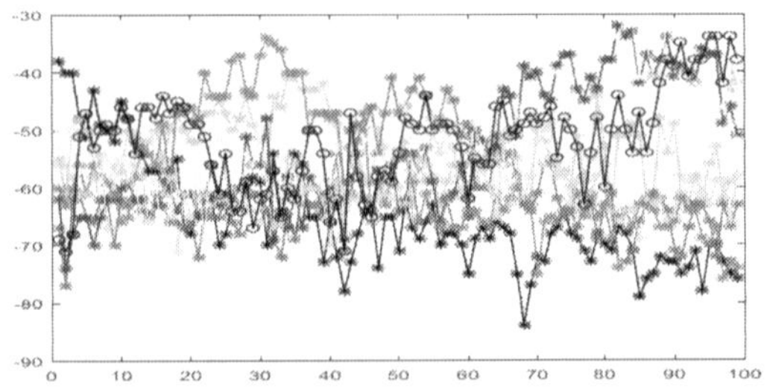

Fig. 2. The eight WAPs RSS values along the survey path.

Moreover, to understand how the proposed integration approach is working, three different experimental scenarios are conducted. The first scenario is to conduct an experiment to use the WAPs RSS values of the dataset. The second scenario is to test an experiment to use only the collected magnetometer measurements of the dataset. The last scenario is to integrate both WAPs RSS values and magnetometer measurements of the dataset. Note, the proposed approach is also implemented when the measured WAPs RSS values and magnetometer measurements values with fingerprint information are available to demonstrate the validity of the algorithm performance. The process of matching between offline WAPs RSS values and online WAPs RSS values is handled by the algorithm using the k-NN technique. As described in previous section, the position of the smartphone is estimated using equation 3.

$$X_{es} = \frac{\sum_{i}^{k} Xi}{K}\ ,\ Y_{es} = \frac{\sum_{i}^{k} Yi}{K} \quad (3)$$

Where k is the number of selected records and x and y is the coordinates of the smartphone position in the pre-surveyed map and Yes and Xes is the estimated position. Note, the positioning accuracy via the root mean square error (RMSE) is used to evaluate the performance of the experimental scenarios, as indicated in equation (4).

$$pos_{err} = \sqrt{\frac{1}{2}(X_{es} - X_t)^2 + (Y_{es} - Y_t)^2} \quad (4)$$

Where pos_{err} is the positioning accuracy, X_{es} and Y_{es} are the estimated X and Y location coordinates, and X_t and $Y_t$are truth values of X and Y location coordinates respectively.

4 Results and Discussions

For the sake of the performance of the proposed integration, each technology (including WiFi and magnetometer) has its own strengths and weaknesses based on its unique properties. For example, WAPs signal readings and magnetometer measurements can provide location information based on different positioning techniques. However, the primary challenge with using magnetometer sensor measurements in indoor environments is the issue of interference caused by building structures or steel obstructions. Also, WAPs signals will be affected by exiting complex indoors structure such as: thickness of walls, doors, and ceiling. In addition to that, the movable object including people moving in the vicinity will affected the WAPs signal. In contrast to the limitations of a single technology-based localization system, it may take the advantages of their strengths by integrating the technology measurements into the fingerprinting technique based on their qualities. First, this study is going to through how each technology measurements in the proposed positioning approach provides positioning accuracy.

From the experiments, with the first scenario, the WAPs RSS-based fingerprint technique can estimate smartphone location by using eight WAPs RSS values from eight access points. This is followed by comparing them to the RSS value of a test sample. The average positioning error is 1.6164 meters. With the second scenario, when only magnetometer measurements including (x, y and z) are used, the average error of the positioning accuracy is up to 3.4600 meters . This means that the obtained positioning accuracy via WAPs RSS values is better than of using magnetometer measurements. The third scenario is to test the integration positioning approach by combining the two localization technologies measurements, as it is mentioned earlier. The integration is based on taking the advantages of both WiFi technologies and magnetometer sensors measurements. The average positioning error which has been obtained from combining these two localization technologies is 0.7224 meter that's quite less than using each of them alone-based technology. The cumulative distribute functions (CDF) of the achieved positioning accuracy (positioning error) and the positioning error at each fingerprint of the dataset are shown in figure 4 and figure 5, respectively.

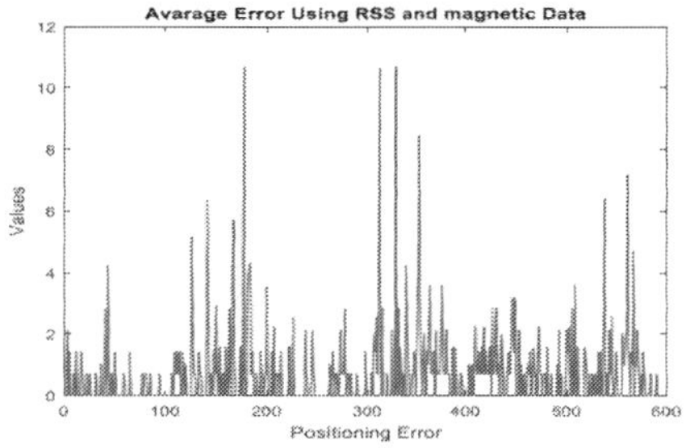

Fig. 5. Positioning error for all the fingerprints of the dataset when both WAPs RSS values and magnetometer is used.

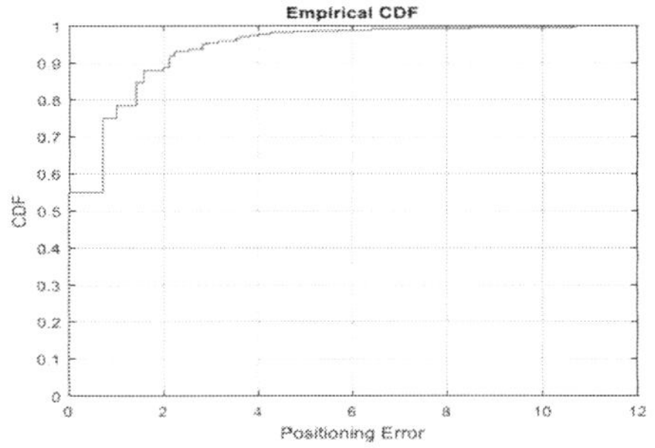

Fig. 4. Cumulative distributions function of the positioning error when both Magnetometer and RSS is used] measurements is used

5 Conclusion

In this study an integration indoor positioning technique using Wi-Fi signals and magnetic field of magnetometer sensor is presented. The integration approach is to improve smartphone position using error minimization and to provide better positioning accuracy. The integration is also compared to fingerprinting technique using WiFi technology and magnetometer measurements separately, as they have possibility of realization in practice. Further, magnetic based navigation system does not need any infrastructure and just

 https://doi.org/10.1051/itmconf/20224201016

requires the magnetic fingerprint map. With this work a set of experiments and different scenarios are conducted to prove the validity of the proposed approach. The experimental results showed that the average positioning errors provided by WiFi RSS values and magnetometer sensors are 1.6662 and 3.74 meters, respectively. Further to effectively localize the position and to improve the positioning accuracy, this study proposed the integration of these two technologies. The experimental results of the integration, shows that better performance can be achieved by, minimizing the smartphone positioning error to 0.7224 meter. Future steps of this study could be based on modifying the k-NN algorithm (as it has been used as a matching algorithm in the fingerprinting technique). The modification is to use feature weighting method to combine RSS values and magnetometer measurements. In addition, some statistical procedures can be applied on the WAPs RSS values and magnetometer measurements to improve the matching process via k-NN or any machine learning algorithms including decision tree, SVM, or deep learning algorithms.

References

1. Ali, M.U., S. Hur, and Y.J.S. Park, Wi-Fi-based effortless indoor positioning system using IoT sensors. 2019. 19(7): p. 1496.
2. Ninh, D.B., et al., An effective random statistical method for Indoor Positioning System using WiFi fingerprinting. 2020. 109: p. 238-248.
3. Mendoza-Silva, G.M., et al., Long-term WiFi fingerprinting dataset for research on robust indoor positioning. 2018. 3(1): p. 3.
4. Ashraf, I., S. Hur, and Y.J.S. Park, Indoor positioning on disparate commercial smartphones using Wi-Fi access points coverage area. 2019. 19(19): p. 4351.
5. Ali, W.H., A.A. Kareem, and M.J.C.U.-E.S.J. Jasim, Survey on wireless indoor positioning systems. 2019. 3(2): p. 42-47.
6. Yeh, S.-C., et al., Study on an indoor positioning system using Earth's magnetic field. 2019. 69(3): p. 865-872.
7. Hanley, D., et al., The Impact of Height on Indoor Positioning With Magnetic Fields. 2021. 70: p. 1-19.
8. Hosseini, K.S., et al. Improving Multi-floor WiFi-based Indoor positioning systems by Fingerprint grouping. in 2021 5th International Conference on Internet of Things and Applications (IoT). 2021. IEEE.
9. Kuang, J., T. Li, and X.J.I.S.J. Niu, Magnetometer bias insensitive magnetic field matching based on pedestrian dead reckoning for smartphone indoor positioning. 2021.
10. Dari, Y.E., S. Suyoto, and P.J.I.J.I.M.T. Pranowo, CAPTURE: A Mobile Based Indoor Positioning System using Wireless Indoor Positioning System. 2018. 12(1): p. 61-72.
11. Keser, S.B., A. Yazici, and S.J.M.I.S. Günal, An F-Score-Weighted Indoor Positioning Algorithm Integrating WiFi and Magnetic Field Fingerprints. 2018. 2018: p. 7950985:1-7950985:8.
12. Pérez-Navarro, A., et al., Challenges of fingerprinting in indoor positioning and navigation, in Geographical and Fingerprinting Data to Create Systems for Indoor Positioning and Indoor/Outdoor Navigation. 2019, Elsevier. p. 1-20.
13. Ashraf, I., et al., A Comprehensive Analysis of Magnetic Field Based Indoor Positioning With Smartphones: Opportunities, Challenges and Practical Limitations. 2020. 8: p. 228548-228571.
14. Gozick, B., et al., Magnetic maps for indoor navigation. 2011. 60(12): p. 3883-3891.

Adaptation for Vigenère Cipher Method for Auto Binary Files Ciphering

Ammar Waysi AlTuhafi[1*]

[1]Department of Computer Science, Knowledge University, Erbil, Iraq

Abstract. Protecting the private information from third party became an important issue in computer science nowadays; cryptography is a science of encryption and decryption information. The first use of cryptography was in ciphering text messages, with the developing of computer usage; other type of files has been appeared such as: picture, audio, video, etc. These types of files can carry private information that needs to be saved from the third part. This paper based on design and implement system for auto encryption/decryption binary files by adapting well-knownVigenère encryption methods, by converting binary files to stream of bytes and encrypting each byte of this stream. The result shows the complete auto protection for the binary files.

1 Introduction

Usually security in computer science can be divided into two main types; network security and information security, the old term of information security usually bounded with texts, whereas, there are numerous number of algorithms that deal with text encryption and decryptions; although most of these algorithms has been proved as perfect algorithms in text encryption, there is no use for these algorithm with another type of data (files) beside text files. Therefore, this study deals with how to use one of these algorithms for encrypt and decrypt binary files [1].

Cryptology is the science where security engineering meets mathematics to make communication mysterious to people except those who are involved in this communication and have the right to read it. Cryptology provides us with tools to do encryption process. The cryptography deals with the secrecy system itself and its design, and cryptanalysis is how breaking this secrecy system [2].

The first use of cryptography was in military and secret agents. In these days we need cryptography everyday because of the daily use of computers and the Internet; and the need to security become personal need beside institutions or organizations, especially when there are commercial transactions over the internet [2].

From the previous, can be concluded, that the first and main use of cryptography is with texts encryption, which is the message between two sides that can be transmitted in secure way because no one can understand the message except the two sides. These days the information

*Correspondingauthor:ammar.waysi@knu.edu.iq, dr.ammar.altuhafi@ieee.org

can be represented in many shapes, such as pictures, video, audio, etc. therefore, encrypt these type of data (files) is required; needing new algorithm to find suitable key for these files is required too, whereas, the key is a basic need in encryption and decryption [3]. Basically every file is a series of bytes in sequence. The value of each byte is between 0 and 255. In general every file is a binary file, but it depends on the data in it, if it contains only text like: letter, numbers, symbols used in writing, then this file can be considered as text file. The binary file is computer readable, not a human readable. All implemented programs are stored in binary files also numeric data files, beside video, audio, picture, etc , in other words text files can be readable, while other file are not[4, 5]. The aim of this research is to design and implement auto cryptosystem for encrypting and decrypting binary files

2 Background

2.1 Vigenère Ciphering Method

This name is back to the Frenchman Blaise de Vigenère, a diplomat who served King Charles IX. Vigenère cipher is a method of encrypting alphabetic by using either the Vigenère square or algebraic way. The Vigenère square is 26X26 table with row heading and column heading from A to Z. The first row of this table has the 26 English letters sequentially, but with starting with the second row to the last one the letter will be shifted one position to the left and the previous letter will moves to the end in cyclic way. i.e. in second row B is shifted to the first position, and A take the last position.

This method the keyword repeated until it equal the length of the plaintext. For example, consider the plaintext is TECHNOLOGY and the keyword is PEN. The keyword repeated as following: TECHNOLOGY
PENPENPENP

To encrypt, take a letter in the plaintext and its corresponding in the keyword, use the keyword letter as row index and plaintext letter as column index, the cross index is the letter in the cipher text. In the example above the first letter is T and its corresponding is P, the cross between them is I which is the first letter in cipher text, repeating this process until all plaintext is encrypt and get the result. The cipher text is: IIPWRBASTN [6].

To decrypt, take the letter in cipher text and its corresponding in the keyword, use the keyword letter and find its row and search in this row for the letter of cipher text, then the head of the column of this letter is the corresponding plaintext letter. So the result of this cipher text IIPWRBASTN is TECHNOLOGY.

To decrypt, take the letter in cipher text and its corresponding in the keyword, use the keyword letter and find its row and search in this row for the letter of cipher text, then the head of the column of this letter is the corresponding plaintext letter. So the result of this cipher text IIPWRBASTN, is TECHNOLOGY.

Vigenère can also represent in algebraic way with the following equation for encryption:

$$C = (P + K)\ Mod\ 26 \quad (1)$$

While used the following equation for decryption:

$$C = (P - K)\ Mod\ 26 \quad (2)$$

While C represents cipher text letter, P represents the letter of plaintext, and K represents the letter of keyword. The letters A-Z are taken as numbers 0-25 so each letter takes his number and implemented in the equations [7].

3 Methodology

The meaning of encryption of binary file is transfer a normal binary file to encrypted binary file, to produce encrypted binary file. No one can understand the contents except the sender and receiver; the following are the steps of encryption with details.

3.1 Key Extraction from Binary File

The key that will be used in the operations will be extracted from the binary file itself, this will be done by choosing two different numbers (K1 and K2), whereas, K1 represents the frequency of repeating the key, while K2 represents the length of the key, i.e. if K1=9 and K2=7, this mean each nine bytes will choose the last byte as a part of the key, this process will continue until the length of the key become seven which is the value of K2; therefore the values of bytes number (9, 18, 27, 36, 45, 54, 63) become the key of Vigenère encryption method.

The key that will be used in the operations will be extracted from the binary file itself, this will be done by choosing two different numbers (K1 and K2), whereas, K1 represents the frequency of repeating the key, while K2 represents the length of the key, i.e. if K1=9 and K2=7, this mean each nine bytes will choose the last byte as a part of the key, this process will continue until the length of the key become seven which is the value of K2; therefore the values of bytes number (9, 18, 27, 36, 45, 54, 63) become the key of Vigenère encryption method, the K1 will be choose at the time when user sends the message, K1and K2 will be extracting from device clock, K1 represent the seconds while K2 represents the milliseconds at the time when user sends the message as said .

3.2 Binary File Encryption by Vigenère

After extracting the key from binary file, the encryption of binary file will happen by Vigenère ciphering method with some modifications to be adaptive, the formula for encryption will be as the following:

$$C= (M + K)\ MOD\ 256 \quad (3)$$

In this operation we add M to K, where M is the number of the bit from the binary file, K is the key that has been extracted, the MOD here is 256 because the ciphering here on binary files not text, whereas the maximum value of the byte is 255. The key's bytes will be inserted in the same places in encrypted binary file.

3.3 Keys Insertion in Encrypted binary File

The K1 and K2 will be sent to the receiver inside file, therefore, the sender and receiver have to know the position of both keys inside the encrypted fil, sometimes can be the first two or last two bytes or any position known for both sides, this insertion will happen before sending the encrypted binary file. Fig.1 shows the methodology of encrypting binary files.

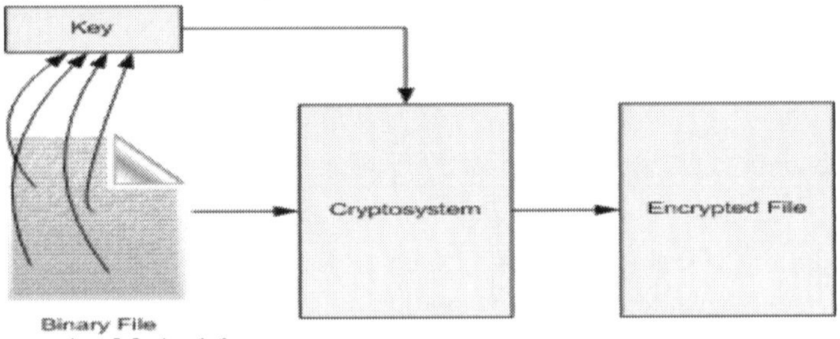

Fig.1. Encryption Methodology

3.4 Decryption Binary Files

Each encrypted item (message or file) need decryption to get it back to original value; the following are the steps of decryption with details.

3.5 Key Extraction from Encrypted Binary File

The method of extracting key from encrypted binary file it is the same method that explains the key extraction in encryption process by using the K1 and K2 which they also extracted from the binary file assaid in section 3.3, based on the position of them in binary file.

3.6 Binary File Decryption by Vigenere

Taking the encrypted file and the extracted key, then apply decryption according to Vigenère method rules by using the following formula:

$$(C - K) = M \quad (4)$$
$$If\ X < 0\ Then\ X = X + 256 \quad (5)$$

In this operation subtract K from C, in case the result in minus (less than zero), the number 256 will be added to get positive number which represent the original value of the byte, after applying this process for the whole file, then the original binary file can be retrieved, fig 2 show decryption of the binary files.

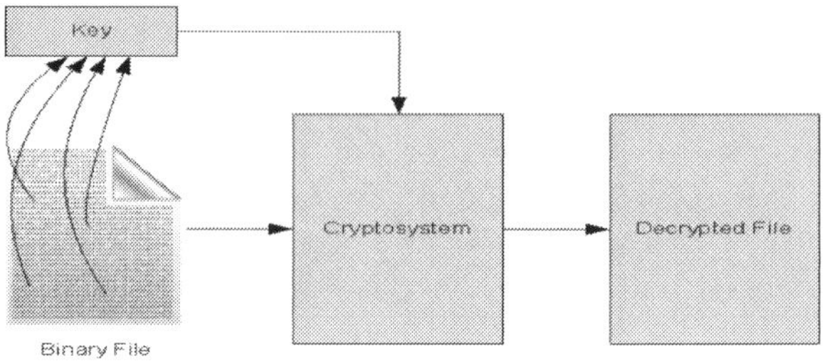

Fig.2. Decryption Methodology

4 Results

The complexity of the algorithm can be written as following: O (m+n+2)

Whereas m is the length of the key that will be extracted from the binary file, n is the length of the binary file because the adapted Vigenère equation will be applied for the whole binary file. The number 2 represents of the K1 and K2; from the previous it is clear that the complexity of the algorithm is low.

Appling this algorithm for on binary files in different types, such as image, video, audio, etc, shows the ability to encrypting the binary files and make software such as photos viewers and audio/video players not able to view or play the file.

Algorithm evaluation against brute-force attack can be calculated according to the number of possible combinations, the following formula shows the possibility of brute-force attack [8].

$$\text{Possible combinations} = C^L \quad (6)$$
Whereas C =possible number of characters and L = password length

The value of C in this algorithm is 256; which is the possible values of a byte because the range of the value of byte is (0-255), while the length of the key can be any number less than file length.

i.e. if the key length is 100, then possible combinations will be as following:

$$\text{possible combinations} = 256^{100} = 6.668*10^{240} \quad (7)$$

To calculate the time that needed to break the key, supposed that an attacker uses China's Tianhe-2 supercomputer, which is the fastest supercomputer. This computer is able to perform $33.86*10^{15}$ floating point operations force per second.

To calculate the time needs to break the key with length 100 byte the following formula can be used.

$$t = \frac{C^L}{33.86\times10^{15}\times(3600\times24\times30\times12)} \quad (8)$$

Applying this formula for key with length 100 byte, the result is $6.33*10^{216}$ years the time that needed to try all the possible combinations [9].

The comparison between the adaption algorithm and well-known Data Encryption Standard (*DES*) ciphering method; DES method cipher depends on bitwise operations which is considered closest ciphering method to this research; DES use key with 56 bit by applying equation 7 for DES, the result shown as following: $6.9*10^{110}$ years the time that needed to try all the possible combinations; which is less $1.1*10^{106}$ years than time needs for the adaptation algorithm [10, 11].

5 Conclusion

This study focus on adaptive encryption method of text messages for binary files. The study shows the ability of encrypting binary files based on Vigenère ciphering method. The key that used to encrypt binary file extracted from the file itself; then this key become a key of Vigenère encryption method, and adding this key to the binary file before sending it. Whereas the binary file acts like text message and combined with the extracted key according to Vigenère encryption method equation, while the sequence of choosing the key parts will be sent as well as inside encrypted binary file. This study shows the ability to encrypt the binary file and disabling any application for reading binary files to open and show its contents.

References

[1] S. Schwartz, "Cryptology for Beginners," *Wissahickon HighAmbler, Pa,* vol. **19002**, (2004).
[2] C. P. Pfleeger and S. L. Pfleeger, *Security in computing*: Prentice Hall Professional Technical Reference, (2002).
[3] R. J. Anderson, *Security engineering: a guide to building dependable distributed systems*: John Wiley & Sons, (2010).
[4] G. Szabo. (2016). *What is a text file and what is a binary file?* Available: https://perlmaven.com/what-is-a-text-file
[5] M. Rouse. (2012). *Binary File*. Available: https://whatis.techtarget.com/definition/binary-file
[6] Q.-A. Kester, "A hybrid cryptosystem based on Vigenère cipher and columnar transposition cipher," *arXiv preprint arXiv:**1307.7786**,* (2013).
[7] Q.-A. Kester, "A cryptosystem based on Vigenère cipher with varying key," *International Journal of Advanced Research in Computer Engineering & Technology (IJARCET),* vol. **1**, (2012).
[8] AceBIT. (2018). *Brute-Force Attacks*. Available: https://www.password-depot.com/know-how/brute-force-attacks.htm
[9] F. L. Malallah, *et al.*, "Irreversible Biometric Template Protection by Trigonometric Function," *International Review on Computers and Software (IRECOS),* vol. **11**, (2016).
[10] P. V. Saraswathi and M. Venkatesulu, "A block cipher algorithm for multimedia content protection with random substitution using binary tree traversal," *Journal of Computer Science,* vol. **8**, (2012).
[11] M. Matsui, "Linear cryptanalysis method for DES cipher," in *Workshop on the Theory and Application of of Cryptographic Techniques*, (1993).

 https://doi.org/10.1051/itmconf/20224201018

Powering over single mode fiber by shortening working time of IoT device

Edvin Skaljo[1*], *Aljo* Mujcic[2], *Petr* Munster[3], *Anis* Maslo[4], *Mujo* Hodzic[5], *Muhamed* Duhandzic[6] *and Emir* Karamemedovic[7]

[1]Department for Physics, University of Sarajevo, Zmaja od Bosne 33, Bosnia and Herzegovina
[2]Faculty of Electrical Engineering, University of Tuzla, Bosnia and Herzegovina
[3]Dept. Telecommunications, Brno University of Technology, Brno, Czech Republic
[4]Deparment for investment, BH Telecom Sarajevo, Bosnia and Herzegovina
[5]Polytechnic Faculty, University of Zenica, Bosnia and Herzegovina
[6]Medical school Zenica, Bosnia and Herzegovina
[7]FabLab Sarajevo, Zmaja od Bosne 8, Bosnia and Herzegovina

Abstract. Fiber optics has revolutionized telecommunication with its superior bandwidth and distance it can span. For its use in IoT networks, some of the limiting factors are the high cost of new installations and the need to power the end device by electrical current. The installations are a part of long-term investments, and one can expect this to be an ever-smaller issue as more fibers are installed. Typically, the newly installed cables contain single-mode fibers. There are a lot of reports on transport of power over fiber, however, majority recommend using multi-mode fibers with a large core or double-clad fibers. In our approach, instead of increasing the core of the fiber, we focus on the possibility of shortening the working time of IoT devices, using the existing single-mode fiber for powering. Also, instead of an expensive PV (photo voltaic) cell with small dimensions and a high efficiency, we propose using the commercially available larger PV cells with an air gap between the end of the fiber and the cells. In accordance with our approach, we successfully conducted an experiment.

1 Introduction

Internet of Things (IoT) networks of devices are growing daily and changing our environment. According to the forecast of the company Statista, the number of IoT devices is expected to triple in the next 10 years (from 8.75 billion in 2020 to 25.4 billion in 2030) [1]. Many IoT devices are sensors that usually consume low power as they typically do not need to be awake all of the time. In sleep-mode, sensors draw from 0 to several μA. While awake, the electrical current can go from a few to several hundred of mA. The awake mode can be from one time per several milliseconds to one per day or so. Duration of the awake mode is typically less than 1s [2]. For example, temperature for weather conditions should be sent every hour with a duration of the awake mode of less than 1s. It should be mentioned that, the longer the sleep mode, the longer is the lifetime of the sensor.

* Corresponding author : skaljo@hotmail.com; edvin.skaljo@pmf.unsa.ba

IoT sensors are often connected to the Internet by LoRa, 3/4/5 G, NB-IoT, Sigfox etc. Fiber optics is rarely used due to two main reasons, namely cost and powering. The cost of fiber optics is very high if the system requires new installations. If fiber optics networks are already installed, the share of fiber optics in the cost of the complete IoT network is significantly reduced. Currently a lot of fiber optic cables are installed underground or on poles. Most of them are based on the standard single-mode fiber (SSMF) or some compatible type, for example FTTH networks are based on SSMF and smaller parts on fibers according to ITU Recommendation G.657.

The other drawback, powering over fiber, has been intensively studied. These studies are characterized by proposing use of multimode fibers or other types of fiber with larger core diameter with possible use of batteries as an auxiliary power supply [3][4][5].

2 Powering over Fiber

There are two basic power transmission architectures over fiber optics depicted in Fig. 1:

- Separating fibers for energy transmission and information transmission;
- Common fiber for powering and information.

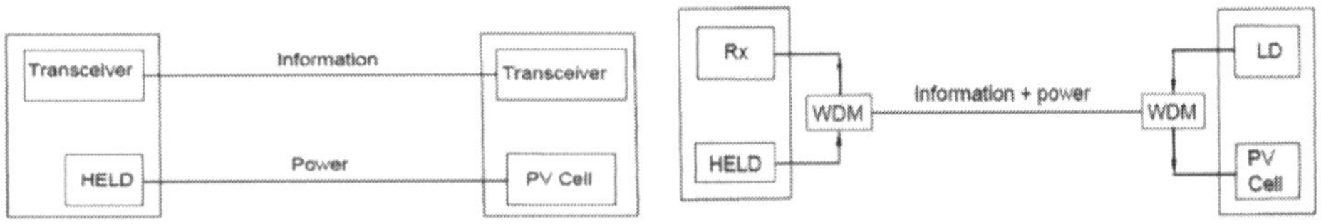

Fig. 1. Power transmmision arcihtecture: (left) two separate fibers for power transmission (right) Common fiber for powering and information flow.

On the sensor side, PV (Photo Voltaic) cell can be used in the following three ways:

1. directly (Fig. 2. a).
2. parallel to a battery (Fig. 2. b);
3. charging a capacitor as the energy storage (Fig. 2. c);

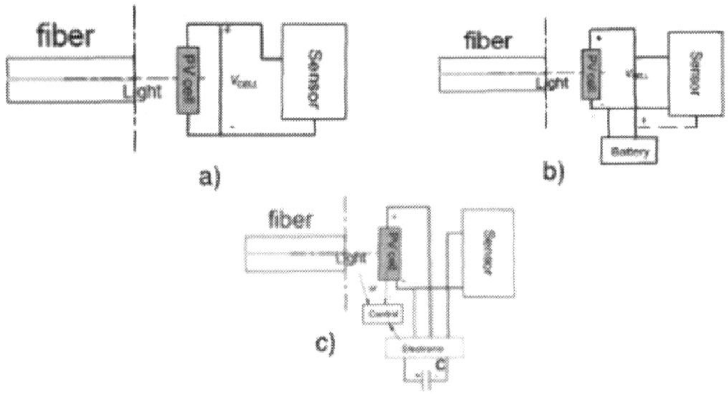

Fig 2. Three cases of receiver: a) Direct mode of PV Cell usage; b) Powering over battery; c) Charging the capacitor

The first case is the simplest but also the most impractical for long-distance applications. It is difficult to achieve sufficient energy for the state of wakefulness at a distance of more than a few hundred meters. In the second case, the transferred energy can be stored in the battery during the sleep mode and consumed during the awake mode. The disadvantage of this model is the need to have a battery, and occasionally replace the battery every few years. The third model has a longer life without intervention because the capacitor lasts significantly

https://doi.org/10.1051/itmconf/20224201018

longer than a battery. However, the current has a variable intensity, so electronic devices must be used to regulate the voltage or current.

2.1 Powering over Fiber - our approach

In our approach, we propose to use power transmission over the same fiber through which the information is sent, according to Fig. 1. right, and the receiver side in according to Fig. 2.

The more detailed approach:

- *SSM fiber* because it is all around us and has a very low attenuation. This allows the development of cost-effective devices for long-distance powering of sensors.
- *Simple PV cells commercially available,* where instead of high field concentrations and high efficiency. PV cells have to be small in dimension with high efficiency to provide sufficient voltage to power sensors from a small surface. In [4], a receiver with 4 diodes in series in a circle with a diameter of 1.5 mm is proposed to achieve a voltage level above 2 V. High efficiency and nonstandard PV cells can cost several times more than the complete sensor.
- *A capacitor for energy storage*. A battery can be used as an energy accumulator, but its lifetime is limited. The lifespan of fiber optics is 25 years or more, electronics as well, so the use of a capacitor is recommended instead of the battery because there is no need for any maintenance for many years of exploitation. The disadvantage of using a capacitor instead of a battery is the variable power supply, namely the capacitor during discharge gives a voltage that decreases exponentially, which requires additional voltage stabilization.
- *The distance between the end of fiber and PV cells* is used to increase the voltage from PV Cell to charge a capacitor. In [6] MPPT (Maximum Power Point Tracking) was used to increase the capacitor charging voltage. In our approach we easily obtain a voltage of 5,6 V for the charging capacitor. The distance from the end of the fiber to the PV cells depends on the numerical aperture of used fiber and the dimensions of the PV cells. In Fig. 3, one can see that the light beam does not effectively cover the square cell, the loss is proportional to the ratio of the areas of the circle and the square. Therefore, the best ratio that kept the radius of the radiation circle between 0.5 and 0.707 was experimentally sought. The PV cells were based on amorphous Si (a-Si).

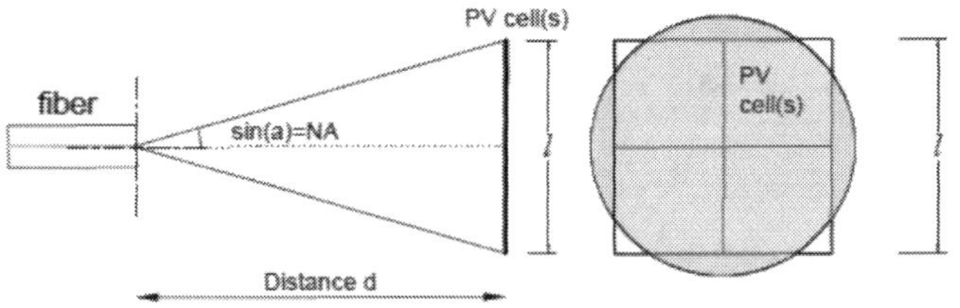

Fig. 3. Fresnel refraction to cover large dimension of PV Cell

- *PV cell based on a-Si*. Due to the limitation of the experiment to a laser wavelength at 650 nm (red laser) with a transmitting power of 10 mW, the choice of PV cell is a-Si. The efficiency of a-Si ranges from 1.9 to 2.4 times less than the efficiency of c-Si [7]. However, c-Si is characterized by a drastic reduction in efficiency - when the radiation power falls below $10 W/m^2$ it is practically unusable. The efficiency of a-Si, on the other hand, remains usable even at values of the radiation density below $0.1 W/m^2$. For comparison, the efficiency of c-Si when the power density is less than 0.1 W/m is below 0.2%, while for a-Si it is about 5% [7]. This phenomenon is crucial in the present

 https://doi.org/10.1051/itmconf/20224201018

approach; it allows us to convert a very small amount of light into electrical power. The choice of a-Si for PV cells makes the power supply system cost-effective.

2.2 Maximum lunched power into the fiber - damage threshold

The maximum optical power that can be inserted into the fiber must be observed through two phenomena: The first is the physical destruction of the fiber core. The second is the effect of light used to transmit energy to the information-carrying signal, in the form of nonlinear effects in the fiber due to the high power of light. The theoretical limit at which there is damage to SSM fibers (SMF-28 Ultra Fiber) is around 870 mW, while for practical use 210 mW is recommended [8]. It should be noted that the power threshold depends on the effective mode area A_{eff} that corresponds to the mode filed diameter (d_{MDF}) and the threshold power density I_{th} [9]:

$$P_{th} = A_{eff} I_{th} = \pi \left(\frac{d_{MFD}}{2}\right)^2 I_{th} \qquad (1)$$

Depending on the specific design, values of 2.5 MW/cm² are mainly used [11].

$$d_{MDF} = 2a(0,65 + \frac{1,619}{V^{3/2}} + \frac{2.879}{V^6}) \qquad (2)$$

3 Experimental setup

The experimental setup consists of the following material, and is sketched in Fig. 4:

Laser diode, wavelength 650 nm, maximum power launched 10 mW, optical fiber, SSM with attenuation at 650nm is 6 dB/km and length 517 m. The distance between fiber tip and PV cells is 15,6 cm. PV cells measure 3 cm x 1 cm and 2.5 cm x 0,9 cm, with output voltage of 5,6 V. Capacitor is electrolyte type with capacitance of 44 μF.

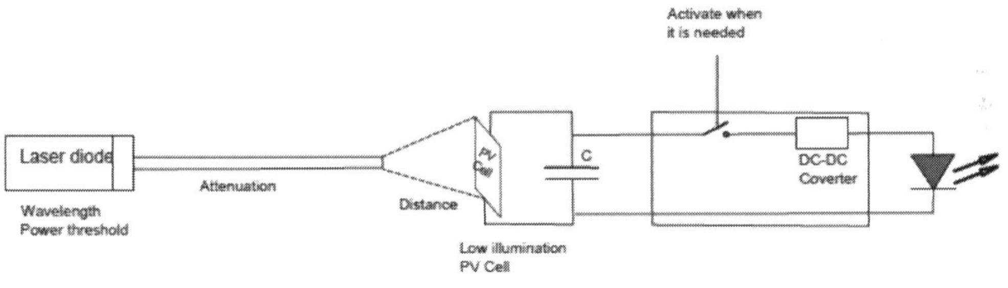

Fig. 4. Experimental setup

Waking up of the device is performed manually, Arduino and voltmeter are used to monitor voltage on the capacitor. A 220 Ohm resistor limits the current and drops the voltage to a green LED (Fig. 5).

Fig. 5. Some figure from laboratory during experiment

4 Discussion and future work

Using inexpensive components including a-Si PV cell and a red 10mW laser, in this experiment it was possible to transfer over 517 m SSM fiber sufficient energy for an LED to emit light for 1s. The current which was obtained was at most around 20 mA. Transferring more power to feed larger load would require charging a larger capacitor with more time between the awake modes. However, at a different, larger wavelength, the attenuation of the fiber decreases, the mode-field diameter increases allowing more power to be coupled to the fiber, making the system more efficient. The main reason for the low efficiency of PV cells on low radiation is the shunt resistance. In [10] authors present CIGS (Copper Indium Gallium Selenide) solar cells on low radiation and different wavelengths. For our work the wavelength of 850 nm could be interesting. In addition to this, a power management unit providing constant voltage at the output can further increase the efficiency. A concentrator reflecting back to PV cell the light that is not absorbed could also increase the efficiency as the refractive index of the cell is rather high.

5 Conclusion.

This is a report on use of inexpensive a-Si PV cells for transferring power over a long distance via a single-mode fiber for powering e.g. low-power sensors. The efficiency of a-Si at very low power densities is demonstrated making it advantageous compared to otherwise better crystalline Si.

Applications of this device are guided by its features such as low price, very long service life and the possibility of placing the device high or low temperature environments. With this in mind, the potential applications may be: Flame sensors in combination with non-combustible optical cables, to obtain information on wildfires; in mines, especially mines with flammable gases; in seismological research; in research with large long-term amounts of snow or underwater sensor network, and generally, wherever sensors are installed in a hostile environment or built into a structure after which the sensor may be difficult or impossible to access.

Acknowledgements

This project is supported by the Government of Canton Sarajevo/ Bosnia and Herzegovina: "Powering of sensors by electrical energy transferred over fiber optics (in Bosnian: "Napajanje senzora električnom energijom prenesenom preko optičkog vlakna").

References

1. *IoT global annual revenue 2019-2030*, https://www.statista.com/statistics, Published by Arne Holst, Aug 25, (2021),
2. Usmonov, Maksudjon, and Francesco Gregoretti, *Design and implementation of a LoRa based wireless control for drip irrigation systems, 2017* 2nd International Conference on Robotics and Automation Engineering *(ICRAE)*. IEEE, (2017).
3. López-Cardona, J. D., et al., *Remote optical powering using fiber optics in hazardous environments,* Journal of Lightwave Technology **36.3** (2018): 748-754.
4. Rosolem, Joao Batista, and R. Roka, *Power-over-fiber applications for telecommunications and for electric utilities, Optical fiber and wireless communications* (2017): 255-278.
5. Wei, Pu, et al., *Power-over-fiber system with energy management scheme for electric power industry*, Optical Engineering **58.9** (2019): 096104.
6. Casaleiro, João Carlos, et al., *On the Feasibility of GPON Fiber Light Energy Harvesting for the Internet of Things*, i-ETC: ISEL Academic Journal of Electronics Telecommunications and Computers **4.1** (2018): 9.
7. Efika Schubert, Markus B., and Jürgen H. Werner, *Flexible solar cells for clothing*, Materials today **9.6** (2006): 42-50.
8. *Laser-Induced Damage in Silica Optical Fibers*: https://www.thorlabs.com, December (2021);
9. Todoroki, Shin-ichi, *Fiber fuse propagation behavior*, INTECH Open Access Publisher, (2012).
10. Swartz, Craig H., et al., *Measurement of shunt resistance and conduction band offset in Cu (In, Ga) Se2 solar cells through joint analysis of temperature and intensity dependence of open-circuit voltage and photoluminescence*, Japanese Journal of Applied Physics **59.5** (2020): 055504.

AUTHOR INDEX

EDP Sciences – Web of Conferences
17, Avenue du Hoggar
Parc d'Activité de Courtabouf
BP 112
F-91944 Les Ulis Cedex A
France

ISBN 978-1-7138-4453-2

Fourteenth International Conference on Software Engineering Advances (ICSEA 2019)

Held at SoftNet 2019

Valencia, Spain
24 – 28 November 2019

IARIA